Engineering Nature
Art & Consciousness in the Post–Biological Era

Edited by Roy Ascott

intellect
Bristol, UK
Portland, OR, USA

First Published in the UK in 2006 by
Intellect Books, PO Box 862, Bristol BS99 1DE, UK
First Published in the USA in 2006 by
Intellect Books, ISBS, 920 NE 58th Ave. Suite 300, Portland, Oregon 97213-3786, USA
Copyright ©2006 Intellect Ltd

All rights reserved. No part of this publication may be reproduced, stored in a retrieval system, or transmitted, in any form or by any means, electronic, mechanical, photocopying, recording, or otherwise, without written permission.

A catalogue record for this book is available from the British Library

ISBN 1-84150-128-X
Cover Design: Gabriel Solomons
Copy Editor: Wendi Momen

Printed and bound in Great Britain by 4edge, UK.

700.105
ASC

Contents

7 Preface

9 Introduction

1. **The Mind**
11 1.1 Towards a Conscious Art
Robert Pepperell

17 1.2 Effing the Ineffable: An Engineering Approach to Consciousness
Steve Grand

33 1.3 (Re)Constructing (Non)Dualism
Andrea Gaugusch

39 1.4 Another View from the Blender
Michael Punt

47 1.5 Bio–electromagnetism: Discrete Interpretations
Nina Czegledy

53 1.6 MEDIATE: Steps Towards a Self–Organising Interface
Paul Newland, Chris Creed and Maestro Ron Geesin

57 1.7 Facts about P–E–M (Psycho–Enhanced Memberships) you must know
Armando Montilla

65 1.8 Happenstances
Evgenija Demnievska

69 1.9 Ontological Engineering: Connectivity in the Nanofield
Roy Ascott

2. **The Body**
77 2.1 Are the Semi–Living Semi–good or Semi–evil?
Ionat Zurr and Oron Catts

91 2.2 Absent Body Project
Yacov Sharir

97 2.3 Our Body as Primary Knowledge Base
Kjell Yngve Petersen

103 **2.4 Electronic Cruelty**
Gordana Novakovic

109 **2.5 Design Against Nature**
Anthony Crabbe

115 **2.6 Why Look at Artificial Animals?**
Geoff Cox and Adrian Ward

121 **2.7 Biopoetry**
Eduardo Kac

3. The Place

127 **3.1 Real Virtuality: Authenticity in Electronic and Non–electronic Environments**
Eril Baily

133 **3.2 Sharing Virtual Reality Environments across the International Grid (iGrid)**
Margaret Dolinsky

137 **3.3 Interactive, Responsive Environments: a Broader Artistic Context**
Garth Paine

145 **3.4 Towards Defining the 'Atmosphere' and Spatial Meaning of Virtual Environments**
Ioanna Spanou and Dimitris Charitos

153 **3.5 Symbiotic Interactivity in Multisensory Environments**
Stahl Stenslie

159 **3.6 Aesthetics Within Ego Shooter Games**
Maia Engeli

163 **3.7 Creative Communities in Networked Hybrid Spaces**
Mauro Cavalletti

171 **3.8 From Multiuser Environments as Space to Space as a Multiuser Environment: Cell Phones in Art and Public Spaces**
Adriana de Souza e Silva

177 3.9 Arch–OS v1.1 (Architecture Operating Systems), Software for Buildings
Mike Phillips and Chris Speed

183 3.10 (an)Architecture, Eros, Memory: the Naxsmash Project
Christina McPhee

189 3.11 Who Plays the Nightingale?
Claudia Westermann

195 3.12 Breeding, Feeding, Leeching
Shaun Murray

4. The Text

201 4.1 The Potential of Electronic Textuality
Dene Grigar

207 4.2 Art and Information
David Topping

213 4.3 Metaphorical Vestiges on Info–Viz Trails
Donna Cox

217 4.4 Interstellar Messaging, Xenolinguistics, and Consciousness: LiveGlide Meets the SETI Enterprise
Diana Reed Slattery and Charles René Mathis

223 4.5 Art and HCI: A Creative Collaboration
Ernest Edmonds, Linda Candy, Mark Fell and Alastair Weakly

5. The Art

229 5.1 The Idea Becomes a Machine: AI and A–Life in Early British Computer Arts
Paul Brown

235 5.2 The Interactivity of the Moving Document as the Diegetic Space of Consciousness
Clive Myer

241 5.3 Assimilating Consciousness: Strategies in Photographic Practice
Jane Tormey

247 5.4 Simultaneity, Theatre and Consciousness
Daniel Meyer–Dinkgräfe

253   **5.5 Cinematic Soteriology: Darshanic Effects in the Tamil Bakthi Films**
Niranjan Rajah

259   **5.6 Search For Utopia: Human Consciousness and Desire**
Julia Rice

263   **5.7 Super Interactivity: Art, Consciousness, and the Dawn of the Participatory Age**
Alex Shalom Kohav

271   **5.8 Pete and Repeat were Sitting on a Fence: Iteration, Interactive Cognition and an Interactive Design Method**
Ron Wakkary

279   **5.9 Visual Art as an Earning Process: The New Economics of Art**
Nicholas Tresilian

285   **5.10 Culture, Ecology and the Real**
Paul O'Brien

291   **5.11 The Immersive Experience of Osmose and Ephémère: An Audience Study**
Hal Thwaites

299   **5.12 On Making Music with Artificial Life Models**
Eduardo Reck Miranda

305   **5.13 Artistic Strategies for Using the Arts as an Agent through the Creation of Hyper–Reality Situations**
Karin Søndergaard

6. The Future

311   **6.1 The Nanomeme Syndrome: Blurring of Fact and Fiction in the Construction of a New Science**
Jim Gimzewski and Victoria Vesna

329   **Contributor Biographies**

Preface

These papers have been selected in part from the proceedings of the Fifth International Research Conference of *Consciousness Reframed: Art and Cconsciousness in the post–biological era* that took place July 31st –August 3rd, 2003 at the Caerleon campus of the University of Wales, Newport. Since the first conference, which I convened there in 1996, a consistent level of excellence, originality and insight has been maintained in the papers presented –a clear indication, I believe, that the issue of consciousness in the arts, and the significance of science and technology in this context, are of dynamic relevance to contemporary culture. Other papers, which add significantly to this issue, have been selected from Technoetic Arts, the international journal of speculative research, published by Intellect Ltd, of which I am founding editor.

The issues signalled in these articles, together with the proceedings of previous *Consciousness Reframed* conferences, and the texts of the journal, provide a rich exploration of the technoetic principle in art (that which prioritises the technologies of consciousness), and, taken as a whole, constitute a valuable archive of the ideas that are informing emergent fields of transdisciplinary theory and practice. Many perspectives can be brought to bear on art in the post–biological era, and whether they are scientific, poetic, spiritual, ethical, or social, they each occupy a place at the cutting edge of our 21st century's artistic adventure.

Matters of mind, and the navigation, and perhaps eventually the explanation of consciousness are of cardinal significance to both art and science, just as they have always been at the very centre of the search for knowledge and our exploration of the numinous in previous, even archaic cultures. This book embodies the writings of artists and scholars from twelve countries in four continents, and so may be considered as a valuable reflection of international thought, practice and meditation on the place of technology and consciousness research in the current techno–culture.

Roy Ascott
Director of the Planetary Collegium, University of Plymouth

Introduction

Past ontologies have always taken the view that nature engineers events, rather in the way that the cosmos engineers time and space. Simply put, there is either a big design with a built–in telos or an emergent coherence that can be rationally apprehended. Classically, we have assumed that time flows only one way, events are connected in causal relationships, reality has been engineered *ab initio*. The clockwork model persists, with many cognitive scientists seeing the mind as an epiphenomenon of the brain. With this model it is possible to conceive of reverse engineering nature, the better to surpass or at least mimic natural process.

Quantum science, while recognising the linear lawfulness of macroscopic events, takes a different view: at levels below its atomic and molecular materiality, nature is without the status of certainty; the complimentarity of particle and wave, matter and spirit, provides only for indeterminate outcomes. Non locality and non linearity are qualities at the very foundations of life, the specificity of all quantum events depending on our active participation as observers.

The artist, dealing mostly in metaphor, is sensitive to these conflicting descriptions of nature, while depending finally on intuition to resolve or enrich the ambiguities of interpretation. As one who builds worlds with images, texts, movement or sound, the artist is particularly responsive to the idea that nature is constructed. Some may go so far as to see the material world as an epiphenomenon of the mind, or as an aspect of the flux and flow of a universal field of consciousness.

The understanding of what constitutes nature has undergone considerable revision in the past hundred years, with some arguing that the term should be removed from enlightened discourse, and others asserting that it was never more than a ploy for enforcing the immutability of moral values, a part of the project to normalise the irrational and uncontrollable in human behaviour. In some cultures it replaces the spiritual .

Computer assisted technologies have allowed us to look deeper into matter and out into space, to elicit or construct meaningful patterns, rhythms, cycles, correspondences, interrelationships and dependencies at all levels. Computational systems have led us to an understanding of how the design and construction of our world could constitute an emergent process, replacing the old top–down approach with a bottom–up methodology. Nano science has been particularly suggestive in this, as well as other, even more challenging respects. Telematic systems have enabled us to distribute ourselves over multiple locations, to diversify our identity, to extend our reach over formidable distances with formidable speed. We have learned that everything is connected, and we are busy in the technological process of connecting everything. Nature and consciousness may be terms to describe the inward and outward manifestations of a profound, complex and numinous connectivity that not simply pervades the universe but constitutes it.

The role of mixed reality technologies and telematic media can be seen to have much potency in creating a dynamic equilibrium between natural and artificial processes and systems, which is to say, between the given and the engineered. New insights into the nature of consciousness may result in new forms of art or equally may return us to the poetic metaphors of traditional cultures or arcane traditions. Without doubt, definitions of nature, as with those attempting to explain consciousness, are negotiable. The writings in this collection provide valuable and varied contributions to these negotiations. They are grouped simply in those categories most associated with nature and consciousness that variously affirm or redefine their reality: *Body*, *Mind*, and *Place*, along with the attributes of *Text* and *Art*. The book ends, as it begins and tracks throughout, with the *Future*!

Roy Ascott.

1. The Mind
1.1 Towards a Conscious Art

Robert Pepperell

The goal of a scientific study of human consciousness seems to be perpetually thwarted by a logical dilemma. Whereas conventional science proceeds through empirical observation of an extrinsic object by a scientific subject (for example, the mineral deposit is observed by the geologist), in the case of self–consciousness it is less clear how one person could objectively observe someone else's inner experience. Despite the range of methodological tools available to experimenters, including introspective reports and sophisticated scanners, they are inevitably left with a rather second–hand picture of what is going on within the minds of their subjects, which are also their objects of study. Because the subject and the object become thus entangled in attempts to observe inner experience, it seems we might never be able to represent the self–conscious mind with anything other than itself. Without a way of representing, or even visualising subjective experience it could remain immune to scientific scrutiny.

In an attempt to address this I propose that the concept of infinite regression, which is normally associated with the 'homuncular fallacy', be reinterpreted productively, in a way that puts self–reference at the heart of our conception of phenomenal experience. Infinite regression is a recurring motif in consciousness studies and is usually treated with suspicion at best and derision at worst. If the various sensory data we draw from the world is bound together somewhere in the brain and observed by an internal 'self' or 'pilot' to whom we can attribute the experience of existence, we run the danger of supposing a further pilot in order to account for the first pilot's experience and so on. The problem is clearly illustrated in the following image, which suggests by its own analogy that to see the car on the miniature screen inside the subject's head the clipboard–holding observers (suitably attired in white coats) should carry identical apparatus in their own heads of which they themselves are part.

The potential absurdity of this line of explanation has lead many thinkers to see infinite regression as sterile and to propose ways of describing consciousness that avoid invoking it (for example, Dennett 1991). But the fact that we bump up against infinite regression so frequently when trying to understand the self–conscious mind may suggest there is more at stake than a simple logical error.

I would suggest that there are in fact two kinds of infinite regression, which can broadly be categorised as conceptual and physical. Conceptual regressions, such as

Figure 1

Aristotle's 'Unmoved Mover' or the homuncular fallacy referred to above, are logically irresolvable. Physical regressions on the other hand, such as mirrors that reflect each other or cameras that see what they are recording, do not suffer the same logical flaw since the constraints of natural law demand that any self–referential physical system must reach some sort of resolution. So in the case of the self–reflecting mirror the very nature of light and the way it reacts to certain surfaces sustains a condition of infinite regression without leading to a conceptual black hole. One could say much the same about video feedback, which occurs when a video camera is suitably directed at a monitor displaying the camera's output (the classic paper on the subject is (Crutchfield 1984). Here a number of non–linear attributes, such as screen discretisation, changing light levels and minute voltage variations, can produce an overall state of great complexity, beauty and variety from what is essentially an infinitely regressive process (see Web link in bibliography for examples).

The idea that consciousness may be linked to self–reference –that is, something looking at itself looking at itself ad infinitum –has a very long history, particularly in Eastern philosophy. In his book *Zen Training*, Katsuki Sekida (1985) outlines a theory of immediate consciousness using the behaviour of mental actions called 'nen', approximately translated from Japanese as 'thought impulses'. For the purposes of this paper I want simply to sketch the basic principle of nen–action, introduced by a passage from the book itself:

> *Man thinks unconsciously. Man thinks and acts without noticing. When he thinks. 'It is fine today,' he is aware of the weather but not of his own thought. It is the reflecting action of consciousness that comes immediately after the thought that makes him aware of his own thinking . . . By this reflecting action of consciousness, man comes to know what is going on in his mind, and that he has a mind; and he recognises his own being.*

According to Sekida, thought impulses rise up all the time in our subconscious mind, swarming about behind the scenes, jostling for their moment of attention on the 'stage'. Of these 'first nen', as Sekida calls them, most go unnoticed and sink back into the obscurity of the subconscious, perhaps to return later in some harmful form. But those that are noticed, just momentarily, by the *reflecting* action of consciousness (the 'second nen') form part of a reflexive sequence that supports our sense of self–awareness. The second nen follows the first so quickly they seem to occur simultaneously –they seem to be one thought. The obvious problem of how we know anything of this second nen is resolved by the action of a third nen which 'illuminates and reflects upon the immediately preceding nen' but 'also does not know anything about itself. What will become aware of it is another reflecting action of consciousness that immediately follows in turn' and so on. Meanwhile, new first nen are constantly appearing and demanding the attention of the second nen.

For the sake of simplicity this sequence is initially presented as a linear progression but Sekida goes on to elaborate the schema with a more subtle, matrix–like organisation while the basic principle remains. What follows from this is that, as Sekida states, 'Man thinks unconsciously'; there is no localisation of conscious thought, no conscious object as such, other than an ongoing, infinitely regressive loop of self–reflections. Nevertheless, because of the rapid sequencing of the internal reflections, one has the impression of a sensible self, much in the way that one has the impression of moving objects in the cinematic apparatus. This theory would suggest that the notion of the 'self' does not exist outside a process of continuous self–reflection, nor in any part of that process.

Analogies between video feedback and consciousness

It was Sigmund Freud who, as far back as 1900 in *The Interpretation of Dreams*, suggested consciousness acted as a 'sense organ for the perception of psychical qualities' (Freud 1976) –in other words, as a 'sixth' sense. It does seem that self–consciousness allows us to see what we see, hear what we hear, taste what we taste, etc. Bearing this in mind (and referring back to the larger project of constructing a 'conscious art' system), I'd like to propose the following analogy between a video feedback system and a conscious being. Imagine an extended video feedback system that includes a video mixer that merges four distinct sources. The monitor not only displays the image from a camera –A –but also a 'mix' of three other sources, or sub–signals –B, C and D –such that all four sources merge in the monitor display and are observed by the camera. In such a set–up the feedback image generated by the looped signal A also incorporates the information from the sub–signals B, C and D. Whichever is the strongest sub–signal tends to have a greater influence on the overall properties of the feedback image. From this quite simple set–up one can draw surprisingly rich analogies with the operation of human consciousness.

As many have observed, consciousness is always consciousness of *something* and the content of our awareness is derived from one of three sources: objects in the world apprehended by the sense organs, sensations from inside the body (e.g. pains, tingles, hunger), mental data (e.g. ideas, memories, thoughts) or combinations thereof. Activity in the world, body and brain can go on quite happily without us being in any way conscious of it but something special or 'phenomenal' occurs when we do become aware of it. For the purposes of this analogy, think of the video camera as the agent of self–reflection that not only sees

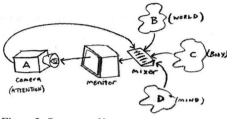

Figure 2: Sensory self–awareness

the combined data–sources of world B, body C and unconscious mind D (the 'content' selected for prominence by signal strength, or level of excitation) but sees 'itself' seeing them, insofar as its own signal A is fed back into what it 'sees' (by analogy Freud's 'sixth sense'). As the signals flow an overall unitary state is reached that obviates the necessity for any further unifying agent, or homunculi, as the system is self–generating whilst also being infinitely regressive, or self–referential. The same principle might be applied to any reflexive sense: the smelling of smells, for example, or the feeling of feelings. Multiplied over the whole sensory system, one might start to speculate how a system with a capacity for self–awareness of some kind might emerge through having an integrated array of feedbacking self–sensors.

Perceiving continuity and discontinuity
Elsewhere I have argued that nature is neither inherently unified nor fragmented but that the human sensory apparatus gives rise to perceptions that make the world seem either unified or fragmented to differing degrees depending on what is sensed (Pepperell 2003). The process of infinite regression in video feedback illustrates how a complex, non–linear self–referential system can spontaneously give rise to patterns of similarity and difference. If consciousness is in any way analogous to video feedback it may help us to understand why, in a world that may be neither inherently continuous nor discrete, we are able to experience both qualities.

The binding problem
Some of the functional parts in the video feedback system are necessarily non–local (the camera lens must be a certain distance from the monitor) but are also connected by light or, like the brain, electrical conduits. In the case of video feedback, non–local components can give rise to coherent global behaviour that cannot be isolated to any part of the system. However, the feedback effect itself can only be observed locally; that is, on the monitor or in the camera eyepiece, despite the distributed nature of the system. Whatever the confusion or variation might be between the sources or sub–signals B, C and D in the analogy described above, the overall feedback image will retain a certain stability and unitary coherence as long as all the variables stay within certain parameters. This could be likened to the unitary coherence of first–person experience –the so–called 'binding problem'.

The value of these analogies for the scientific study of consciousness is the way in which video feedback can practically represent the complexity that can emerge from a physical process that is essentially regressive. It may serve, therefore, as a means of analogically visualising what consciousness might 'look' like and thereby offering a way of observing the phenomenon.

Conscious art

I would like to sketch out, in very general terms, how these ideas might inform a practical investigation into the production of a self–conscious work of art.

It might be interesting for the reader to know that the ideas presented here originated less from the relevant philosophical literature than from a combination of introspection, personal experience and artistic enquiry. In particular, the practice of meditation and examination of its related philosophies have helped to clarify a number of issues to do with the behaviour of the mind and its relation to the body and the world. In addition, whilst using LSD some years ago I experienced vivid recursive patterns of luminous colour very similar to those seen in video feedback, which triggered an intuition about the self–referential operation of the visual system and by extension the mind. It is these experiences, together with the various pieces of interactive art I have produced and exhibited over the years, that have circuitously led me consider how it might be possible to construct an object of art that displays some self–awareness.

Using some of the principles discussed above, a system is envisaged that combines three sources of data from 1) the external world (with sensors for light, sound, and pressure, etc.), 2) the internal state of the system (such as levels of energy and rates of information flow, etc.) and 3) repositories of images, sounds and texts to be activated by rules of association (what one might describe, crudely, as 'memories'). These three data sources will be synthesised into an overall system state, which is then 'observed' by separate sub–system of sensors, much like the video feedback referred to above. This observed state is then fed back into the overall system state and re–observed, indefinitely. In this way the system will generate a condition of infinite regress not dissimilar to that found in video feedback, which it is envisaged will achieve some overall coherence. At the same time, because conditions will constantly vary in the exhibition space (in terms of audience actions, internal system data states and associative links with stored data), the global behaviour of the system will be non–linear and unpredictable.

However, I should stress that I am not claiming any such system, even if it performed well, would actually be conscious in the same way that we are. Nor am I even claiming it would be quasi–conscious, or yet further, that it would be an accurate model of how conscious processes occur in humans. To claim any of these would not only pre–empt the results of the investigation but would suggest a far grander purpose than the thesis I have presented here could justify. At best the system might have a rudimentary functional self–awareness.

But even given the obvious limitations, I do expect many more artists to become interested in the creative possibilities of self–aware systems. This is on the basis that such systems will have unique and compelling qualities, including a capacity for producing semantic richness in response to audience behaviour, at the same time as generating a *frisson* of expectation amongst audiences as they apprehend an artwork that displays, albeit in the mildest of forms, some of the same sentient behaviour they recognise in themselves.

Conclusion

Infinite regression then, understood in relation to phenomenal experience, may be understood as a process of perpetual self–reference of self–observation, however this might occur within the physical substrate of the human system, the non–linear nature of which can give rise to intricate and novel behaviour. By exploiting the mechanical and analogical properties of video feedback systems, including their inherent complexity and creativity, one can envisage a functional model of a self–referential system that might inform a wider theory of a conscious, or self–aware art.

(This paper is based on a larger text published in Technoetic Arts *Volume 1:2 2003).*

Bibliography

Crutchfield, J. (1984). 'Space–Time Dynamics in Video Feedback'. *Physica*, 10D, pp. 229–45.

Dennett, D. (1991). *Consciousness Explained.* London: Penguin.

Freud, S. (1976). *The Interpretation of Dreams*. London: Pelican.

Pepperell, R. (2003). *The Posthuman Condition: Consciousness Beyond the Brain*. Bristol: Intellect Books.

Sekida, K. (1985). *Zen Training: Methods and Philosophy*. New York: Weatherhill.

www.videofeedback.dk/World/

1.2 Effing the Ineffable: An Engineering Approach to Consciousness

Steve Grand

The easy problem of 'as'

According to the *Shorter Oxford English Dictionary,* the word 'ineffable' has a number of meanings, one of which seems, inexplicably, to have something to do with trousers. More usually, it's used to describe a name that must not be uttered, as in the true name of God, but it can also mean something that is *incapable* of being described in words. Consciousness, in the sense of what it really means to be a subjective 'me', rather than an objective 'it', might turn out to be just such a phenomenon. Language, and indeed mathematics, may be fundamentally insufficient or inappropriate as explanatory tools for describing consciousness, and Chalmers' 'Hard Problem' (Chalmers 1995) might in the end prove quite ineffable.

Nevertheless, there are other means to achieve understanding besides abstract symbolic description. After all, even apparently trivial concepts are often inaccessible to verbal explanation. I am completely unable to define the simple word 'as' and yet I can use it accurately in novel contexts and interpret its use by others, thereby demonstrating my understanding operationally.

Linguistically speaking, an operational understanding involves the ability to use or interpret a word or expression in a wide variety of contexts. However, for an engineer, an operational understanding is demonstrated by the ability to build something from first principles and make it work. If I can build a working internal combustion engine without slavishly copying an existing example, then I can be said to understand it, even if I am incapable of expressing this understanding in words or equations.

An operational definition of consciousness might exist if we built an artificial brain (attached to a body, so that it could perceive, act and learn), which reasonably sceptical people were willing to regard as conscious (no less willing than they are to ascribe consciousness to other human beings).

Of course, simply *emulating* consciousness, by building a sophisticated puppet designed to fool the observer, should not be sufficient. Here we see a danger inherent in the Turing Test (Turing 1950), since it turns out to be quite easy to fool most of the people for at least some of the time. For an artificial brain to show that we have obtained an operational understanding of consciousness, it should thus be built according to a rational set of *principles*, accessible to observers, and implicitly demonstrate how those principles give rise to the behaviour that is regarded as conscious. In other words, it must encapsulate a theory at some level, even if the totality of that theory cannot be abstracted from its implementation and described in words or simple equations. In particular, the conscious–seeming behaviour must *emerge* from the mechanism of the machine, rather than be explicitly implemented in it, just as a model of an internal combustion engine must drive itself, not be turned by a hidden electric motor.

I shall return to the distinction between emulation and replication later. For now, I simply want to assert that synthesis is a genuine, powerful method for achieving understanding and a respectable alternative to analysis, despite its low status in the reductionist toolkit of science.

Strictly, synthesis of an apparently conscious machine would only constitute an answer to Chalmers' 'Easy Problems' –a description of the mechanism by which consciousness arises, rather than an explanation of the nature of subjectivity. Nevertheless, it may be as close as one can get and it may still demonstrate a higher understanding in an operational sense, in a way that a written theory cannot do.

Attempts to synthesise consciousness might at least offer an answer to a 'Fairly Hard Problem', lying somewhere between those set by Chalmers. Theories of consciousness (or at least heartfelt beliefs about it) tend to lie at the extremes of the organisational scale. To some, including me, consciousness is a product of high levels of organisation: an emergent phenomenon that perhaps did not exist anywhere in the universe until relatively recently and could not exist at all until the necessary precursors had arisen. To me, a conscious mind is a perfectly real entity in its own right, not some kind of illusion, but is nevertheless an invention of a creative universe. Yet for many, consciousness belongs at the very opposite end of the scale: something fundamental –something inherent in the very fabric of the universe. Overt Cartesian dualism lies here, as does the fictional notion of a 'life force' and Penrose's rather more sophisticated retreat into the magic of quantum mechanics (Penrose 1989). Consciousness is clearly not a property of medium levels of organisation but at which of these two extremes does it lie?

To me, the 'fundamentalists' don't have a leg to stand on, for simple philosophical reasons. Just ask them whether they think their elemental Cartesian goo is simple or complex in structure. If complex, then it is surely a machine, since consciousness is a property of its *organisation*, not its substance. But why add the complication of a whole new class of non–physical machines when we haven't yet determined the limits of physical ones? Alternatively, if they think their fundamental force or elemental field is unstructured and shapeless, as Descartes envisaged it, how do they explain its complex properties? How can a shapeless, sizeless, timeless nothingness vary from place to place and moment to moment?

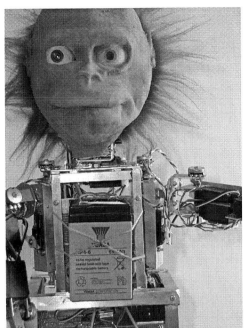

Why does it give rise to intricate behaviour? A force can push but it can't choose to push one minute, pull the next and then do something quite different. Moreover, if consciousness resides in some universal goo, why did it have to wait for the evolution of complex brains before its effects became apparent? Why does it require such specialised (and delicate) equipment?

Despite its absurdity, many people cling on to essentially Cartesian views, even if they don't admit to it, but successfully creating an artificial consciousness would tip the balance in favour of the emergent viewpoint, since no fairy dust would have been added to achieve the result. Failure, of course, would prove nothing.

Figure 1: Lucy the Robot MkI poses for a passport photo

Returning to the more practical but still extremely important and challenging 'Easy Problem' of the mechanism that underlies conscious thought, I contend that synthesis might not simply be a useful alternative to analysis but may prove to be the *only* way in which we might understand the brain. The brain may be genuinely irreducible: incapable of abstraction into a simple model without losing something essential. Neuroscience tries to understand the brain by taking it to pieces, and although this is a vital part of any attempt at understanding, at its worst, it is like trying to understand how a car works by melting one down and examining the resulting pool of liquid metal. It isn't steel but the *arrangement* of steel that makes a car go –and the principles that underlie this arrangement. Taking complex non–linear systems to pieces to see how they work leaves one with a lot of trees but no sign of the wood.

It may be that the brain has no basic operating principles at all, of course. To many observers it is an amazingly complicated hotchpotch –an *ad hoc* structure patched together by evolution, in which each functional unit has its own unique way of doing things. Explicitly or implicitly, many also regard the brain as a hardwired structure (despite overwhelming evidence to the contrary) and their focus is therefore on the wiring, rather than the self–organising processes that give rise to it. In truth, the latter may be relatively simple even though the former is undoubtedly complex.

This attitude to the undeniable complexity of the brain reminds me strongly of the way people interpreted the physical universe before Isaac Newton. The way that planets move in their orbits seemed to have nothing in common with the way heat flows through a material or a bird stays in the air –each required its own unique explanation. And then Newton made three simple observations: if you push something it will move until something stops it; the harder you push it, the faster it will go; and when you kick something your toe hurts because the thing kicks back just as hard. These three laws of motion brought dramatic coherence to the world. Suddenly (or at least inexorably), many aspects of the physical universe became amenable to the same basic explanation. I suggest that one day someone will uncover similar basic operating principles for the brain and, as their relevance emerges, so the brain's manifold and apparently disparate, inexplicable properties will start to make sense.

I should emphasise that the above statement is *not* intended as a defence of ultra–reductionism. The existence of elegant, fundamental principles of operation does not 'explain away' the brain or consciousness –the principles are absolutely no substitute for their implementation.

I submit that trying to build a brain from scratch is more likely to uncover these basic operating principles than taking brains to pieces or studying vision in isolation from hearing or cortex in isolation from thalamus. As an engineer, one is directly faced with the fundamental questions –the basic architecture must come first and the details follow on. To start with the fine details and try to work backwards towards their common ground can quickly become futile, as the facts proliferate and the need for specialisation causes people's descriptive grammars to diverge.

Deep simulation

One of the best reasons to use an engineering approach, and especially computer simulation, as a means to understand the brain is that computers can't be bluffed. Philosophers can wave their hands, psychologists can devise abstract theories that appear to fit the facts but such explanations require human interpretation and people are extremely good at skipping lightly over apparently small details that would crush a theory if they were given sufficient attention. Take the instructions on a packet of shampoo: 'wet hair, massage shampoo into scalp, rinse off and repeat'. The author of these instructions was presumably quite satisfied with their accuracy and sufficiency and millions of people have since read them and believed they were acting exactly as instructed. Yet give the instructions to a computer and it would never get out of the shower, since they contain an infinite loop. Humans know that they should only repeat the instructions once and don't even notice that the 'theory' embedded in the algorithm is incomplete. Computers, on the other hand, do exactly what they are told, even if this is not what you thought you were telling them.

If computer synthesis is ever to succeed in demonstrating that consciousness is an emergent phenomenon with a mechanistic origin, or even if it is only to shed light on the operation of natural brains, certain mistakes and misapprehensions need to be addressed. I've met at least one prominent neuroscientist who flatly refuses to accept that artificial intelligence is possible, owing to an insufficiently deep conception of the nature of computers and simulation. Several others accept that computer simulation is useful for gaining insights into brain function but insist that, no matter how sophisticated the simulation, it can't ever *really* be conscious and can never be more than a pale imitation of the thing it purports to represent. There are good and bad ways to approach computer modelling of the mind and, unfortunately, many people take the bad route, provoking well-deserved, but unnecessarily sweeping, criticism.

I should therefore like to differentiate between two kinds or levels of simulation, 'shallow' and 'deep', because I think many people's objections lie in an unwitting conflation of the two. But to begin with, I need to make some assertions about the nature of reality.

Intuitively, we tend to divide the universe up into substance and form –tangible and intangible. Electrons and pot plants are things –hardware –but paintings and computer programs are form –software. Moreover, we tend to assign a higher status to hardware, believing that solid, substantial things are real, while intangible 'things' are somehow not. Such a pejorative attitude is deeply embedded in our language but I suggest that this distinction is completely mistaken: substance is a kind of form; the intangible is no less real than the tangible; hardware is actually a subset of software.

Take electrons, for example. We intuitively think of electrons as little dots, sitting on top of space. But really an electron is a propagating *disturbance* in an electromagnetic field. It persists because it has the property of self-maintenance. If you were to create a set of randomly shaped 'dents' in the electromagnetic field, many of them would lack this property and rapidly fade away or change form but electron-shaped dents would remain.

So an electron is not a dot lying *on* space, it is a self-preserving dimple or vibration *of* space –it is form, not substance. And if electrons, protons and neutrons are form, then so is a table and so, in fact, are you. The entire so-called 'physical' universe is composed of different classes of form, each of which has some property that enables it to persist. A living creature is a somewhat higher level of form than the atoms of which it is painted. It even has a degree of independence from them –a

Figure 2: Lucy MkII begins to take shape. The object in the centre is her right arm

creature is not a physical object in any trivial sense at all: it is a coherent pattern in space through which atoms merely flow (as they are eaten, become part of the creature and are eventually excreted).

I submit that everything is form and that simple forms become the pigment and canvas upon which higher persistent forms are later able to paint themselves. The universe is an inventive place and once it had discovered electrons and protons, it became possible for molecules to exist. Molecules allowed the universe to discover life and life eventually discovered consciousness. Conscious minds are now a stable and persistent class of pattern in the universe, of fundamentally the same nature as all other phenomena. Perhaps there are patterns at even higher levels of organisation, of which we are as oblivious as neurons are of our minds.

Computers are allegedly the epitome of the distinction between hardware and software but the dichotomy blurs when you consider that the so-called hardware of a computer is actually a pattern drawn in impurities on a silicon chip. Even if you regard the physical silicon as true hardware, the computer itself is really software. But leaving aside metaphysics and starting from the first level of software –the machine instructions –computers are a wonderful example of an endless hierarchy of form. At the most trivial level, instructions clump together into subroutines, which give rise to modules, which constitute applications and so on. The particular class of hierarchy I have in mind is subtly different from this one and has a crucial break in it but the rough parallel between the hierarchy of form in the 'real' universe and that inside the software of a computer should be apparent.

Consider this thought experiment: suppose we took an extremely powerful computer and programmed it with a model of atomic theory. Admittedly there are gaps in our knowledge but our theories are complete enough to predict and explain the existence of higher levels of form, such as molecules and their interactions, so let's assume that our model is sufficiently faithful for the purpose. The computer code itself simply executes the rules that represent the generic properties of electrons, protons and neutrons. To simulate the behaviour of *specific* particles, we need to provide the model with some data, representing their mass, charge and spin, their initial positions and speeds, and hence their relationships to one another. If we were to provide data about the electron and proton in a hydrogen atom and repeat this data set to make several such atoms, we would expect these simulated atoms to spontaneously rearrange themselves into hydrogen molecules (H2), purely as a consequence of the mathematics encoded in the computer program. If we were then to add the data describing some oxygen molecules and give some of them sufficient kinetic energy to start an explosion, we would expect to end up with virtual water.

The computer knows nothing of water. It only knows how electrons shift in their orbitals. But if the database of particles and their relationships was large enough and our initial model contained the rules for gravity as well as electromagnetism, then the simulated water should flow. Water droplets will form a cloud and rain will fall. If we add trillions of virtual silicate molecules (rock) then the rain will form rivers and the rivers will cut valleys. Nothing in the code represents the concept of a valley –it is an emergent consequence of the data we fed into our simulation.

Given a powerful enough computer and a hypothetical scanner capable of extracting the necessary configuration data from real objects in our world, we would expect to be able to scan the atoms of a physical clock, feed the numbers into the simulation, and watch the resultant virtual clock tick. Now imagine what would happen if I turned the scanner on myself. Unless the Dualists are right, it seems to me that I would find myself copied into the simulation. One of me would remain outside, marvelling at what had happened, while the other would be startled to find itself inside a world of virtual clocks and rivers but otherwise perfectly convinced that it was the same person who, a moment earlier, was standing in the computer room. The virtual me would strenuously assert that it was alive and conscious and who are we to disbelieve it?

Many people do. They insist that the virtual me would just *look* as if it thought it was conscious. It wouldn't be real, they say, because it isn't really made of atoms, it is only numbers. But what evidence is there for this conclusion? I accept that the electrons and protons are not real but merely mathematical models that behave *as if* they were particles. But the molecules, the rivers, the clocks and the people are emergent forms, which have arisen in exactly the same way that 'real' atoms, rivers and clocks arise. What's the difference? An emergent phenomenon is a direct consequence of the interaction of components having certain properties and arranged in a certain relationship with one another. It is the properties and the relationships that give rise to the phenomenon, not the components. To my mind, molecules that arise out of the properties and relationships of simulated particles are just as much molecules as those that arise from so–called real particles.

Whether you accept this metaphysical argument about reality doesn't much matter. I partly wanted to draw a distinction between code–driven and data–driven programming. In our thought experiment, the simulation became more and more sophisticated, not by adding more rules, but by adding more data –more relationships. Indeed the total quantity of data hasn't really increased, just its complexity. In the initial state, the program consisted of a few rules describing atomic theory, and a vast database of particle positions and types, all of which were initially set to zero. To add more sophistication to the model, we simply made some of these values non–zero. We might equally have started out with a database filled entirely with random numbers, in which case I would expect the system to develop of its own accord, discovering new, stable, molecular configurations and losing those that are not. Eventually, such a model might even discover life. Data–driven programming has a distinct bearing on computational functionalism in AI and lies at the heart of many people's objections to the idea that computers can be intelligent and/or conscious.

My assertion is that while computers themselves cannot be intelligent or conscious, they can create an environment in which intelligence and consciousness can exist. Trying to program a computer to behave intelligently is practically a contradiction in terms –blindly following rules is not normally regarded as intelligent behaviour. But if those rules are not rules for intelligent behaviour but rules for *atomic* behaviour (or more practically, neural and biochemical behaviour), then genuine intelligence, agency and autonomy can arise just as they can arise from the equally deterministic blind rules that govern 'real' atoms or neurons. To think otherwise is dualism.

Many AI researchers try to build computational models of intelligence in the shallow, explicit way that I rejected above. Psychologists develop computational models of intelligence, too, and these models are generally abstract, symbolic and relatively simple. Such abstract models may well be helpful as *theories* of intelligence but they don't constitute intelligence itself. If you were to embed an abstract psychological model of the mind directly into a computer, you would merely have *emulated* the behaviour of the mind as a shallow simulation and it would almost certainly fail to capture all of the mind's features. A real mind emerges out of myriad non–linear interactions, born of the relationships between the properties of its neurons and neuro–modulator molecules. There may be severe limits on how much this can be simplified. No matter how convenient it might be to describe the brain's basic operating principles in linguistic or mathematical terms (as I will myself later), this does not mean that the abstraction is itself sufficient to implement a mind. An elephant is an animal with a trunk like a snake, ears like fans and legs like tree trunks. But strapping a cobra and a pair of fans to four logs doesn't make an elephant.

To recap: a shallow simulation is a first–order mathematical model of something. It is only an approximation and merely a sham –something that behaves *as if* it were the system being modelled. But deep simulations are far more powerful and, I suggest, have a different metaphysical status. Deep simulations are built out of shallow simulations. They are second–order constructs and their properties are emergent (even if they were fully anticipated and deliberately invoked by their designer). Some phenomena *have* to be emergent and mere abstractions of them are no better than a photograph.

To create artificial consciousness, I suspect we can't just build an abstraction of the brain. We have to build a real brain –something whose implementation level of description is at least one step lower than that of the behaviour we wish it to generate. Nevertheless, it can still legitimately be a *virtual* brain because (unless Penrose is right) we don't need to descend as far as the quantum level, only the neural one. Computers are wonderful machines for creating virtual spaces in which such hierarchical structures may be built. Trying to make computers themselves intelligent is futile but trying to use computers as a space in which to implement brains from which intelligence emerges is not.

Everybody needs somebody

If synthesis might be more powerful than analysis at extracting the core operating principles of the brain and the computer might be capable of turning an implementation of these principles into something from which a mind emerges, where on earth do we start actually developing such a model?

First, brains simply don't work in the absence of bodies. Absurdly futile attempts have been made to imbue computers with such things as natural language understanding, without providing any mechanism through which words might carry meaning. It's all very well telling a computer that 'warm' means 'moderately high temperature' but unless temperature actually has some consequence to the computer, such symbols never find themselves grounded. Bodies and the world in which they are situated provide the grounding for concept hierarchies.

Also, intelligence (if not consciousness) has more to do with learning to walk and learning to see than it does with playing chess and chattering. People often assume that getting a computer to see is a simple problem but this is because they are only consciously aware of the *products* of their visual system, not of the raw materials it has to deal with. Point a video camera at a scene and examine an oscilloscope trace of the camera's output instead of the image itself and suddenly the problem of interpretation seems a lot harder. A creature with no sensory perception would be unlikely to develop consciousness at all. Moreover, when we consciously imagine a visual scene we do so by utilising our brain's visual system. Vision, and other such 'primitive' aspects of the brain, are absolutely essential components of the conscious mind and therefore need to be understood and replicated.

Finally, to get at the brain's fundamental operating principles we need to tackle as many aspects of brain function as we can at once, otherwise we will fail to spot the common level of description that unites them. In any case, interaction between multiple subsystems is important for many aspects of learning. For instance, developing binocular vision may rely on the ability to reach out and touch things to calibrate estimates of distance, while at the same time the capacity to reach out and touch things may depend on having binocular vision.

For all these reasons, the best synthetic approach to understanding the brain is probably to build a robot. This robot should be as comprehensive as possible, to allow for multi–modal interaction and the exploration of many aspects of perception, cognition and action. In my case I decided to build a humanoid (actually, anthropoid) robot, which I named Lucy (Figure 1).

One important thing to mention is the extent to which I've tried to give Lucy biologically valid sensors and actuators. Brains evolved to receive signals from retinas and talk to muscles. They didn't evolve to listen to video signals and talk to electric motors. The signals leaving the retina are not a bit like a video picture –they are distorted, blurred, and represent contrast ratios rather than brightness, amongst other things. These differences between biology and technology might be extremely important. If we are to understand how the brain functions, it is crucial that we try to speak to it in its own language. For that reason, much of Lucy's computer power is devoted to 'biologifying' her various sensors and actuators.

Castles in the air

Given a suitable robot and the appropriate computing methodology, we're just left with the small challenge of figuring out how the brain works. How do we do that, exactly?

My own approach is to simultaneously work from both ends of the problem. Neuroscience and psychology have uncovered a dazzling array of symptoms of brain activity, some detailed, some abstract, all bizarre. All of these phenomena must contain clues about the underlying mechanism and the problem is essentially one of finding the unifying level of description. For example, it is strange enough that schizophrenics hear voices inside their heads but who would have guessed that many can also tickle themselves? The two seem to be quite unconnected. Yet, if you regard both as disorders of self/non–self determination, everything starts to make sense. The voices in a schizophrenic's head are presumably his own thoughts but the vital mechanism that normally masks such self–generated perceptions to prevent them from being confused with external stimuli has failed. Exactly the same logic explains perfectly why most of us are unable to tickle ourselves while schizophrenics can. It is by trying to find such unified levels of description that we

can hope to grasp the essence of what is going on underneath and see glimpses of the core mechanisms at work.

Simultaneously, one can begin from first principles and a blank sheet of paper and work upwards. What does a brain have to do? What key engineering problems does it have to face? For instance, one problem is the significant time it takes signals to travel from the senses through to the deeper parts of the brain, especially when considerable processing is involved. This observation alone is enough to suggest a raft of possible mechanisms and consequent predictions, upon which evidence from neuroscience can be brought to bear. In fact, signal delays are the mainstay of my developing theories of the brain and why I think mammalian brains have evolved in the way that they have.

Figure 3: Lucy learning to recognise the orientation of visual edges. The righthand screen shows two views into a portion of her primary visual cortex

To use this approach on such an incredibly hard problem requires relying on skyhooks to hold up otherwise unsupportable hypotheses long enough to see where they might lead. Progress means following up flaky ideas, stimulated by dubious and possibly absurd interpretations of the evidence. For instance, a whole chunk of my work grew out of a silly line of reasoning about why chemically staining one part of the cortex makes it appear blobby while another area looks stripy. My logic is almost certainly complete nonsense but it led me on a fruitful path towards a practical mechanism for visual invariance –an otherwise highly mysterious phenomenon.

For this reason I find myself reluctant to record any of my 'findings' in a technical context yet, since so many of my ideas are still suspended by skyhooks. This article is meant to be no more than a position paper on behalf of the synthetic approach, not an exposition of a theory of the brain. Nevertheless, it's worth giving a general overview of the picture of Lucy's brain that is gradually emerging, since it demonstrates the synthetic method in action (figure 3), and also shows how counter–establishment an idea can become when it is allowed to follow its own nose. The degree to which Lucy's brain mirrors natural brains remains to be seen, but all of my ideas are backed up by evidence from neuroscience to a greater or lesser extent, and most have been implemented and shown to work inside the real robot, at least up to a point.

There is no stereotypical model of the brain –no specific theory to which all neuroscientists currently subscribe. Nevertheless, there is a distinct underlying *paradigm* – a set of assumptions and models that are held more or less unwittingly, more or less vociferously by many cognitive scientists and lay people. For the sake of comparison with Lucy's rather dissident brain architecture, this paradigm can be shamelessly caricatured as follows:

The paradigmatic brain is a factory, in which raw materials flow in from the senses at one end, are modified and assembled into intentions in the middle, before squirting out the other as motor signals. Very little traffic of any significance passes in the other direction. The factory is a fairly passive structure – more like a maze of empty corridors through which nerve signals pass than a room full of noisy conversations that continue even in the absence of input. It is very complicated and has many highly specialised departments, each with its own unique way of doing things. These specialised structures are hardwired by evolution to perform operations on the data (as opposed to being consequences of the data themselves). The nerve activity that runs through them is presumed to be finely detailed and precise, more like the signals in a computer than ripples on a wobbly jelly. The system is also taken to be largely reactive, whereby programmed actions are triggered only by incoming sensory signals – any apparently pre-emptive action is dismissed as illusory. In this paradigm, work on the 'higher' aspects of thought must wait until all of the complex lower details have been fully understood and the 'C word' is not to be uttered for at least another century.

Lucy's brain, on the other hand, is a responsive balancing act – a dynamic equilibrium between two opposing forces, which I've christened Yin and Yang. These two signals flow in opposite directions along different pathways in her brain, meeting at many points along the way, and playing an important role in several distinct aspects of her brain's activity and development. Her virtual brain begins as a simple repeated structure, onto which a mixture of sensory experience and regular, internally generated 'test card' signals write a story of ever-increasing complexity and specialisation. The specialisation in her cerebral cortex is not hardwired but dynamically created and maintained, much as frail ripples in the sand are defined and held firmly in place by subtle regularities in the movement of waves. This map of specialised areas is essentially a mirror of the world outside – a model, or anti-model, of the world, whose deepest purpose is to undo all the changes that the world wreaks upon Lucy, just as a photographic negative placed over its own positive turns the whole image back to a uniform grey.

Like our own cerebral cortex, Lucy's is divided into many specialised regions, which have pulled themselves up by their own bootstraps. Nerve activity dances on the surface of these regions in sweeping patterns, the shapes of which carry the information. Each dance takes place within a specific coordinate frame: retinal or body coordinates towards the periphery, and more shifting, abstract spaces deeper in. In Lucy these coordinate frames are (or will be eventually) entirely self–organised. The yin and yang streams carry with them information about other coordinate systems through which they have passed and the tension between these flows generates a morphing effect, creating new intermediate frames between the fixed peripheral ones. Some of this morphing occurs during development but it needs to be periodically maintained and eventually this will happen while Lucy sleeps –an alternation of slow–wave sleep with dream sleep is a crucial component of her self–maintenance. I suspect that this basic mechanism for stopping brains from falling apart is more fundamental, evolutionarily speaking, than the more sophisticated mental processes that it makes possible, so perhaps the Australian aborigines were right: perhaps consciousness arose out of the Dreamtime and we dreamed long before we were awake.

Unlike the largely unidirectional, pipelined structure of the stereotypical brain, Lucy's neural circuitry is arranged like an inverted tree, with both sensory and motor signals arising from the same leaves. Yin flows in from the leaves towards the trunk, while yang flows outwards to the leaves. Yin invokes yang, and yang invokes yin, sparking off each other like lightning dancing among clouds. Sensory signals flow inwards in search of something to connect with –something that can make sense of them. Meanwhile, expectation, attention and intention flow outward, looking for confirmation that they are on the right track or are achieving their aims. Each affects the other.

The system is in a perpetual state of tension between sensory data on the one hand, telling Lucy what is happening (or, since nerve signals travel slowly, what has just happened), and an opposing set of hypotheses or intentions on the other, which provide a context for the interpretation of these sensory signals and proclaim what Lucy expects (or intends) to happen next. The outgoing yang signals constitute a model of the world, forming a set of predictions on different timescales. Incoming yin confirms or denies, pushes and pummels this interpretation,= and meanwhile the predictions fill in gaps or compensate for delays in the sensory data, creating a running narrative of what is going on and allowing Lucy's mental world to remain one step ahead of the outside world.

Some of these yang signals can be described as attention, some as expectation and some as intention. What you call them depends only on their context: they are fundamentally the same. If a yang signal erupts at a leaf that has control of eye movements, then one might regard this as an attentional signal, swivelling the eyes to focus on a potentially interesting or dangerous stimulus. But in other circumstances it could be viewed as an expectation, turning the eyes to where something is expected to appear next. In both cases we could reinterpret the signal controlling the eyes as an intention to move them –essentially a prediction about where they will soon be pointing.

The fundamental purpose of this interplay is to maintain Lucy in a contented and comfortable state. If her needs or the outside world changes, then these flows of yin and yang seek to rebalance themselves, sometimes by changing Lucy's internal beliefs but usually by changing the world outside. At any moment there will be two distinct but related states in Lucy's brain, one represented by millions of yin signals (her sensory state) and the other represented by equal numbers of yang signals (her mental state). The system learns to react to keep these two states in balance. Lucy's mental state, her internal narrative about the world, is what I choose to call her *imagination*. In humans, this is where consciousness resides.

A perfect match between Lucy's mental and sensory states is nigh on impossible. Usually her mental state is trying to keep one step ahead of her sensory state, making predictions or issuing intentions according to context and meanwhile her beliefs may or may not match reality. Different parts of the two systems may be in greater or lesser correspondence. If incoming yin signals are straightforward enough to be able to trigger corresponding yang responses right out at the leaves, without requiring more advanced contextual processing, then the trunk of the tree does not become involved. But since it can never be silent, it freewheels. In a person, I would say they were daydreaming or thinking.

Sometimes daydreams lead to decisions, and decisions to actions, and in this case yang signals from the trunk would stretch right out to the leaves and cause change —something humans might regard as a voluntary, conscious decision, as opposed to a simpler, unconscious response. The outgoing yang pathway is a limited resource. Because of the very broad and diffuse way that Lucy's nerve signals pass around her cortex (crucial to the way they perform computations), it is only possible for one 'thought' at a time to traverse outwards along any given yang path. So when such a yang cascade occupies one of the central trunk routes, Lucy's whole body is essentially given over to the same thought and no others at that level are possible. Signals that reach right up to the trunk are thus 'conscious' because they take over most of her mind, while those that cause more peripheral activity can happen unconsciously, leaving the rest of the neural tree to float in a daydream world of its own or deal with something else.

At the moment, not all of these ideas have been fully implemented, and there is obviously a stupendous amount left to be understood before I can see how to grow a whole brain. Lucy's brain has only a hundred thousand neurons anyway –a millionth the size of a human brain. Moreover, despite having developed some really strong intuitions about the basic operating principles that should allow her brain to self–organise from scratch, I'm not yet able to define these rules completely enough that all the messy details sort themselves out, without me having to intervene. I accept that it is probably well beyond the wit of one person to solve all of these problems but for the moment things are going well and I intend to keep up this absurd hubris, since I think that non–disciplinary, holistically inclined individuals like me might actually have more chance of success than large teams of specialists. It is very hard for ten people, with ten different mindsets, to examine ten completely different aspects of a problem and yet still see the common ground.

Suppose that I could solve all of these problems eventually, perhaps 20 years from now. Suppose that I could implement the rules for how my yin and yang circuits interact, in such a way that a complete, neurologically plausible brain develops entirely of its own accord. Will Lucy then be conscious? I simply don't know. All I can say is that I might have demonstrated an operational understanding of a machine that grows in such a way that it comes to behave *as if* it were conscious. That may have to be enough.

Meanwhile, work continues ...

This research was funded in part by the National Endowment for Science, Technology and the Arts (NESTA).

Further information on Lucy the robot can be found in Growing Up with Lucy: How to Build an Intelligent Android in Twenty Easy Steps *(Weidenfeld and Nicholson 2004).*

Bibliography

Chalmers, D. (1995). 'Facing up to the Problem of Consciousness'. *Journal of Consciousness Studies*, vol. 2, no. 3, pp. 200–19.

Penrose, R. (1989). *The Emperor's New Mind*. New York: Oxford University Press.

Turing, A.M. (1950). 'Computing Machinery and Intelligence'. *Mind*, vol. LIX, no. 236, pp. 433–60.

1.3 (Re)Constructing (Non)Dualism

Andrea Gaugusch

The 'binary theory of the sign' and beyond

Presuming a subject, a sign and an object that is signified using signs, we are moving on the rather beaten track of 'classical semiotics' (see fig. 1). Couldn't it be possible, however, that we only perceive 'static objects' we finally believe to 'signify' because we have constructed them as such 'static objects' in the first place?

One could call the idea of 'labelling' an object using language 'iconic–theoretical thinking' or a 'dualistic way of thinking' or a 'naive realistic way of thinking'. Realism means here that the world is presumed to be meaningfully 'formed' independently from the act of cognising it as such a meaningfully 'formed' world. Let's consider, for example, a diagrammatic cross–section of the vertebrate eye (see e.g. Pessoa et al. 1998, p. 724). Pessoa et al. presume the existence of an 'eye'. They have already perceived/cognised an 'eye' as 'eye', they have already separated the 'eye' from the body, have pictured it on a piece of paper and have finally signified it as 'eye'. Just as the precondition for perceiving a blind spot is –again –a drawing or photo that pictures a blind spot. The knowledge of possessing a blind spot seems to be inseparably united with somebody else, who has already distinguished the 'blind spot' as 'blind spot' from the 'rest'. Considering these interactive games, the so–called 'triad' between a 'blind spot' (an object) and my (the interpretant's) sign 'blind spot' (see fig. 1) begs a great number of questions. Among them:

1) If we are putting name tags on pictures, who put the name tag 'blind spot/nervous system etc.' on a 'blind spot/nervous system' at the beginning of this game?

2) How do we know that these slips of paper are correctly attached to the 'forms'?

3) Why, if literally rooted in a universal experience of nature, would not all languages be the same?

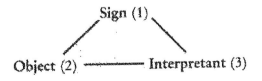

Figure 1: The sign–object–interpretant triad

Picturing our own 'blind spot', we are obviously in the domain of literacy, for the letters forming the words 'blind' and 'spot' are without doubt written down and the picture of a retina is drawn on a piece of paper. And once we have pictured these structures/words, we draw a link between 'sign' and 'picture'. Our 'eye' is, however, still embedded in a body and only under this precondition does the notion 'eye' or 'blind spot' make sense. It seems, therefore, as if we fill in the 'body' in order to talk meaningfully about a 'blind spot', as it is pictured on a piece of paper. But if we do require a body in order to possess a 'blind spot' and if we require somebody else, who points towards our blind spot, and if, finally, the 'blind spot' didn't introduce itself as such, we can conclude: The knowledge of possessing a 'blind spot' requires intersubjective pointing as well as the utterance (or written down words) 'blind spot'. The 'blind spot' as we finally perceive it on a piece of paper goes hand in hand with somebody who has separated this 'spot' from the rest of the 'eye'. And this concept ('blind spot') seems to require at least a second person, already able to utter 'blind spot', pointing towards a specific structure –forming it as a meaningful 'spot'. The realistic idea that there are 'blind spots' out there that we merely 'label' using 'names' seems to go hand in hand with learning reading and writing, which is the precondition for taking 'eyes' out of their context, putting them on a piece of paper, the precondition for perceiving a 'blind spot' as a 'word', as well as the precondition for drawing lines between a 'word' and a 'blind spot'.

These linguistic domains, the domain of orality, meaning that we utter a sequence (without writing it down) and point at an object or touch it simultaneously and the domain of literacy, meaning that we write down 'words', like the words 'blind spot' (on a piece of paper, sand, engraved in stones, on the skin, etc.) and link it to a picture, tend to be intermingled. Reading through recent literature on language acquisition, we will constantly find the notion that children are 'labelling' their environment, labelling 'given forms' out there, as if they learn to attach 'name tags' to given objects/forms (e.g. Bloom 2000 p. 89ff).

This 'labelling idea' evokes, as already stated, numerous questions and carries a fundamental logical error. If a structure like a 'blind spot' does not introduce itself as such, stating, 'Hi, I am your blind spot', the knowledge of perceiving a blind spot as 'blind spot' must necessarily come from a different source than the so–called 'blind spot' out there. And the only possible source is we humans, who point towards aspects and state 'blind spot' –inventing the concept as well as forming the 'object', drawing differences between 'blind spot' and 'no blind spot'. As soon as this game has started, we perceive blind spots as 'blind spots' but if we forget our own game we might separate –later in development –the picture from the name. Or we might become linguists, raising the question of how 'names' and 'objects' are related, overlooking that we have formed any 'object' as 'object' as soon as we state 'object', overlooking that perceiving 'names' as separated from 'objects' is already a human construction –and so is the so–called 'relation' between 'sign' and 'object', as pictured in figure 1.

Psychological sidestep

Piaget (1997) describes children as 'nominal realists', meaning that they do not separate between a 'name' and the 'object being named' until a certain age. The age of 'separation' between 'names' and 'objects' is, however, a matter of discussion (Bloom 2000 p. 178f). It would not be surprising if literacy played a critical role here. Literacy makes a 'word' appear as 'word', whereas on the level of spoken language a word is still 'invisible' and can therefore not be 'attached' to an object (how do you attach something that is not palpable?).

Deepthi Kamawar and Bruce D. Homer (1998) elaborate children's metalinguistic understanding of words and names as follows: 'The traditional view on this issue [the understanding of the concepts of 'word' and 'name', emphasis Gaugusch] is that children begin as 'nominal realists'; that is, they believe that the name for an object is a part of the object being named. According to this view, it is only through general cognitive development that children become nominalists; that is, they come to see that the name is not part of the object, but a particular symbolic or semiotic entity. Recent research suggests that this is far from a linear or monodirectional process . . . The results from this study (Homer et al. 1998) supported the view that there is not a direct and linear progression from infantile realism to nominalist thinking, and that literacy is an important factor in this process' (Jean Piaget Society 1998).

In other words, the idea (of most grown–ups, as well as Piaget) that 'names' do not 'form' benamed objects but are mere signifiers signifying 'given forms' seems to go hand in hand with perceiving 'names' as separated things. The precondition for perceiving 'names' out there is the possibility to 'place them' out there –using 'written language'.

The dualistic worldview separating a priori between 'names' and 'world' dates back to Plato. In the dialogue *Cratylos* he addresses the question of whether things have a proper name 'by nature' (Cratylos) or whether the proper names are a result of conventions (Hermogenes). It is, however, without question that there is a relation between 'objects' and 'names' and therefore the question of how they are connected is emerging. We are, on the contrary, concerned about why/how we separate between 'something' and the linguistic game/name 'something' at all.

We are not, therefore, producing a footnote to Plato here but are trying to show that there is a dualistic precondition in his writings. We are, at least in western philosophy, caught in our dualistic way of thinking, separating a priori between 'objects' and their 'names' –drawing/constructing 'links' and 'lines' between them.

Epistemological conclusions

When constructivists talk about the 'blind spot' of the 'observer' (Maturana and Varela 1984 S. 21ff), they are talking about an 'operationally closed nervous system' as if there were such a 'blind spot' or 'nervous system' or 'observer' out there and names were used to 'label' them. What is missing here is, however, a self–reflexive loop, a second–order–cybernetics, an awareness of the fact that the 'observer', as well as the 'blind spot' or the 'nervous–system', is also our construction and is driven by our perceptual and cognitive qualities and limits.

The a priori separation between 'forms' and their 'names' leads to a separation between different worlds. World one is the world a system constructs and world two is 'ultimate reality'. If one states that there is a world 'beyond the one we live in', we have to ask: How do *you* know that? Separating a priori between 'given forms' and their 'names', we are captured in our dualistic way of thinking. We separate without reflecting on this dualism between 'world' and 'names' and therefore have to come up with the idea that the world we live in is not the 'real' one. Showing, on the contrary, that our names are always already in–worlded as soon as we talk about 'whatsoever', showing that we are forming our world as soon as we perceive our world as 'formed', we conclude: *Nothing is hidden.* We are constantly creating our world; we are constantly projecting our thoughts in the world, acting linguistically with each other. We cannot step out of these concepts; we cannot leave our meaningfully formed world behind us, since any statement of this kind presumes the concepts/precepts in advance. But we can choose either to acquire knowledge within our conventionally real reality or to stop any acting, moving –interacting. We are free to stop any kind of interaction, either with ourselves or with other creatures. If we choose the opposite, if we choose to live in our conventionally real reality, we might learn to perceive ourselves as conjurers, constantly generating a world, enjoying its existence, as we are creating it.

The World Generator – The Engine of Desire as an awareness aid

We tend to forget this process of construction. We tend to perceive things as being 'meaningfully formed' by 'themselves', independent of the observer. We tend to perceive ourselves as 'static subjects', independent from our ongoing discourse –a discourse within ourselves as well as with other selves –embedded in an environment. From a non–dualistic viewpoint we live in our linguistically constructed (micro)worlds, which are formed in an even more manifold way, the more manifold our linguistic–games are formed. People tend to overlook this continuous process of (re)construction. They experience the world as 'static' and as a continuous 'return of the eternal equilibrium'. They are unaware of their process of construction, do not recognise that they themselves form their world –any single moment.

The World Generator –The Engine of Desire (Seaman 1996–7)[1] can be used to make people aware of their constant world–creation and to observe in a scientific and artistic manner how our 'linguistic acting', 'meaning' and 'world' are intermingled.

One of the goals of Bill Seaman's project is to present a computer–based platform for the examination of meaning production. Central is 'interactivity' of the vuser (a term coined by Seaman intermingling viewer and user) as well as the 'recontextualisation' of 'language'. Within *The World Generator* we can observe a world arising through permanent linguistic interaction with various so–called 'media–elements'. 'The techno–poetic mechanism enables the vuser to construct individualised virtual worlds in real time, from a series of media–elements [non–textual 3D computer graphic objects, 3D spatial text objects, 2D texts, video digital image stills, digital video–image stills applied as texture maps, short digital video loops, digital video loops applied as texture maps, digital audio of musical compositions, digital audio presented as spoken text] and processes that are housed within the computer–based environment on virtual container-wheels . . . A physical interface table is directly connected to the virtual space. By manipulating a space ball, a track ball and two toggle buttons, the vuser is empowered to navigate, explore, construct, alter, and abstract media–elements within the space. Interactivity with the authored system enables them to construct and navigate complex worlds . . .' (Seaman 1999).[2]

Within Seaman's *World Generator* it is impossible to separate our permanent interaction with the author (pre–authored language–vehicles), our choices (choosing vehicles interacting with ourselves) as well as the simultaneously arising meaningful world. The outcome of this investigation suggests a 'broader linguistics', a linguistics that is aware of the (linguistically) constructed separation between 'signified' and 'signifier'.

We are not merely 'signifying' pre–given objects/forms anymore; rather, we are constantly generating a world, until something emerges as a meaningfully formed 'blind spot', 'nervous–system', or someone as an 'observer' observing himself.

Bibliography

Bloom, P. (2000). *How Children Learn the Meanings of Words.* Cambridge: The MIT Press.

Maturana, H. R. And Varela, F. J. (1984). *Der Baum der Erkenntnis:* Goldmann.

Paper presented at the 28th annual meeting of Jean Piaget Society, Chicago, Illinois, 1998, June. http://lsn.oise.utoronto.ca/Bruce/Rliteracy/Spring98.nsf/pages/kamawar

Pessoa, L., Thompson, E. and Noë, A. (1998). 'Finding out about filling–in: A guide to perceptual completion for visual science and the philosophy of perception'. *Behavioral and Brain Sciences, 21,* pp. 723–802.

Piaget, J. (1997). *Das Weltbild des Kindes.* München: DTV. French original published as: Piaget, L. (1926). La représentation du monde chez l'enfant. F. Alcan.

Seaman, B. (1999). *Recombinant Poetics: Emergent Meaning as Examined and Explored Within a Specific Generative Virtual Environment.* Unpublished doctoral thesis. University of Wales, College Newport.

Notes

1. See also http://faculty.risd.edu/faculty/bseamanweb/web/index.html
2. See also http://www.foundation–langlois.org/e/activites/seaman/Recombinant_Poetics.pdf

1.4 Another View from the Blender

Michael Punt

A view from the blender
It is well understood that urban myths about vanishing hitchhikers, corpses in packages and public nudity, etc. should not be taken at face value. Taken as emergent collaborative stories they appear to be expressions of commonly shared anxieties that, for all their gore, tend to rehearse social catastrophe as coincident with social and bodily horror. They are traded freely in intimate groups and, perhaps most remarkably, both tellers and listeners know that what purports to be truth is a rather unoriginal fiction often heard before (see Brunvand 1983). Whatever else may be debated about these stories and their anthropological, social or ideological function, at the very minimum we can conclude that they do have a function beyond their face value and that their impact is generative.

Biology
There is one such myth, which may or may not have any foundation, that if a living sponge is macerated in a blender eventually it will not only reconstitute itself as a sponge but, thanks to the mitochondria, each individual cell will return to its former position in the structure. An urban myth to be sure. Chimeric cell behaviour is well known and sieving sponges and chick embryos in order to watch their reconstitution is a well–known experiment which, as a concept, never ceases to amaze. However, the reality seldom matches the ideal, and as far as the sponge is concerned, the degree to which it is macerated, and a combination of other conditions, determines precisely what happens. More often than not the result is some ghastly mutation but as with all popular reiterations of science, the errors and simplifications in the telling may be not so much mistakes as reflections of the anxieties towards the ideals that science represents in the popular imagination. As a consequence the sponge in the blender can be seen as a story with an uncertain origin, not perhaps about the wonders of the microbiology of cells but a narrative reconstruction of biological science through the interaction of the technological and the organic in which an essentialism of the human being offers some guarantees.

At stake in the blender is the comprehension of the melange of that ridiculous accident called life. On the one hand, some science increasingly makes claims about biological determinism with a parade of selfish genes and meme theory while, on the other, civilisation as expressed through art, science and technology, aspires to greater self–determination. If biological determinism is a strategy in the arms race with a hostile environment, a race that regards two sexed reproduction as a means that allows for genetic restructuring in defence of parasitic attack, then cultural interaction, invention and intervention are evolutionary outcomes of biological processes. Not surprisingly, in this scheme of things it begins to be undeniable that we are not all born equal and it seems that whatever else may change, as long as there is two sex reproduction this will always be the case. The massive anxiety that the slide to sociobiology and eugenics causes may not be a reaction to an unacceptable social horror but a personal irritation and constant reminder that we are merely objects when at the very worst we thought that we were subjects. Little wonder, then, that we treat biological science (genetic engineering, genetic modification, cloning, etc.) with suspicion, irreverence and poorly informed revisionist thinking in ways that drive many scientists to their Manhattans.

Electricity

What deserves reiteration, however, is that the sorts of narratives of resistance to the hegemony of scientific naturalism do not come from the uniformed imagination of artists but are crucial elements in the history of the very intellectual project that, half a century later, in 1942, made such a spectacular hole in the desert at Los Alamos. Their provenance is not myth, legend or folklore but science itself —or at least science of the late 19th century —when the discussion in scientific circles shifted from material things to forces (such as electricity, magnetism and radio waves). In this scientific inquiry the force of 'mind' was not an fanciful excursion into self determinism but a legitimate and perplexing topic that liberated much thinking about electricity and magnetism and, of course, Uranium 235. In 1865 Edward L. Youmans argued that since energy can be neither created nor destroyed, the force of mind also exerted connective influence. To support his advocacy of the withdrawal of attention from 'the material and sensible, [to] the invisible and supersensuous world', he argued for scientific research into the intercommunication between the seen and the unseen (see Luckhurst 2002 pp. 85–7). From Faraday's work on electricity and magnetism to Crooke's radiant matter and Galton's *Inquiry into the Human Faculty*, the interaction of forces invisible to human perception formed the unbounded universe open to scientific enquiry. Whether looking at the asymmetry between male and female reproductive cells (the former being invariably microscopic in contrast to the latter) or radio waves or electricity, a key topic in the late 19th century lay in the interactive relationship between invisible presences. Consequently the expectation put on scientists at the time was for them to describe the known and to raise reasonable conjecture concerning the unknown.

If one kind of technological display seemed to confirm this it was electricity in all its various manifestations from the stealthy to the spectacular. The scientific and popular fascination with electricity was not simply its pyrotechnics and danger but as the contradiction of the hegemony of classical science. Without the slightest embarrassment, populist texts at the close of the century described electricity as a phenomenon and, at the other end of the intellectual scale, Faraday himself declared that he was unsure if his topic were metaphysical or physical. This rather extended view of the scientific world persisted for much longer than that other counter hegemony, modern science, has been willing to admit. In the introductory chapter of *Electricity and Magnetism* (from 1893) by S.R. Bottone of 'The Collegio del Carmine, Turin', author of *Electrical Instrument–Making for Amateurs* and *A Guide to Electric Lighting,* suggests that:

> *Because we know so little of the real nature of the agencies which we designate by the above names [electricity and magnetism] it is perhaps advisable not to attempt any profound theorization in studying their phenomena, but rather to proceed in the same way that a child learns its mother tongue, by synthesis only: that is by an examination of the facts as they present themselves to our notice leaving all theory aside until we have collected a sufficient number of facts on which to theorize, and always remembering that a theory, until it has stood the test of rigid proof in all directions is but a theory, and only useful in so far as it enables us to group together and explain an intelligent manner the phenomena which are passing before our eyes.*
>
> (Bottone 1893 p. 1)

The distrust of theory as fact expressed here contrasts with the epistemological empowerment of vision and practice. This theme is reiterated in Bottone's later book on radiography (1898) in which human centred vision itself is regarded as unreliable as the absolute proof of interaction. After reminding us that (as any photographer knows) certain silver salts are sensitive to the action of the more highly refrangible rays of light (in common with many other bodies including vegetable matter), he points out that the power of decomposing these salts is shared in a greater or lesser extent by heat, electricity, magnetism and mechanical pressure (Bottone 1898). Bottone recounts the familiar experiments with sparks from Wimshurst machines but then describes his own process of photographing with electricity in which there is no discharge through the sensitive film but merely the passage of inductive electricity that produces an image. No light, no eyes, no smug realism, no representation of Renaissance ideology; instead an ontology of photography in which light is not writing itself but a trace of the ever–present interaction of forces in a way that makes them comprehensible to the open mind. This was metaphysics, magic and science as a single continuous topic of research.

Even more counter–intuitive to rationalists, perhaps, was that discharges in a Crooke's vacuum tube produced light with a different colour which seemed to have the power to attract solid objects such as cotton wool outside the tube. The miniature airtight laboratory, complete with necessary vacuum, produced a version of another world in this world. In this convergence of realities both Lenard and Röntgen noticed that electrical discharges in very high vacua produce an invisible light that reverses the normal orders of opacity and transparency. They also observed that different shaped tubes and different metals on the anodes produce different visions of the world in which the opaque is no longer a given. Although much of this seemed perfectly explicable –the visualisation of the interiors could be confirmed by dissection –fluorescent screens complicated the issue since these allowed rays to pass through them and left an imprint of an object behind them –this was not only counter–intuitive but also counter–hegemonic. Each new turn in radiography seemed to move more towards the phenomenal than the material in every sense. It was a marvel that caught the public imagination and outshone the Cinématographe as a fairground attraction. Tragic deaths and an overbearing rationalism have ensured that the only surviving remnant that celebrates the mystery of it all is in the name: X–ray.

More startling and stellar was the insistence by Marie Skladoska (Curie) that atoms might decay and their transformation into light could be visible as radiated energy. The selfless toiling over ten thousand kilograms of uranium is well known, as is her final reward: one–tenth of a gram of a shining element, radium, that refused to obey the laws of classical 19th century science. Ra 226 (226 times heavier than hydrogen) appeared to owe its existence to a transmutation of matter over a period of at least 5,000 million years. This mere fact in itself not only transformed geology but it also threw new challenges to Christian fundamentalists and the sense of human significance in the order of things. Arguably the two greatest discoveries of the 1890s, ones that transformed lives and perceptions, both pointed to interactive and phenomenal dimensions to the world about us. More particularly the coalition of radium and the Röntgen x–ray tube opened a new field of enquiry into electromagnetic waves that showed that, contrary to Leibnitz's exclusive assertion, nature not only flowed but also jumped. This coexistence of contradictory forces in a single coherent universe opened the way for broader interpretations of the spiritual.

The instrumental confirmation of other dimensions to reality challenged bourgeois reliance on the epistemology of vision and, along with the fascination with X–rays as a popular entertainment in fairgrounds, so too exhibitions of various technologies also spoke to another dimension. The popular engagement with science and technology as an elevating and distracting spectacle in world's fairs, museums and public lectures extended to a fascination with interaction at a distance and, like serious scientific expositions, telepathic, spiritualist and mesmeric practices were often almost indistinguishable from amusing entertainment. As a consequence the distinction between the paranormal and materialist science was not always clear at either the public or the professional level. Nonetheless it soon became apparent to influential members of the British Association for the Advancement of Science that a firm distinction should be made. By the 1880s the 'Society for Psychical Research collated strange mesmeric effects,

Spiritualist mediumship, apparitions and ghostly manifestations, Crooke's 'psychic forces' and Cox's psychism, and reconceived them through its principal organising term: telepathy' (Luckhurst 2002 p. 59). Telepathy became the single conduit in the heterogeneous terrain of psychic manifestations. Throughout the second half of the 19th century it became the dominant manifestation of psychic phenomena to attract the scientific anti–spiritualists as well as the more populist antagonists of superstition such as conjurors.

Despite the teleological claims of modern(ist) historians, the Cinématographe and the Kinetoscope, the X–ray's great fairground rivals, were not enlisted in a documentary project to represent life as we apparently perceived it. On the contrary, many of the early pioneers, including Edison himself, had a well–known fascination with the afterlife. The quite special possibilities moving picture technology offered for transcending the physical world were (famously) noted at the idea stage of the Kinetoscope and, as the first reports of the Cinématographe point out, despite its relatively degraded image quality, this kind of moving picture technology offered an opportunity to be in both this world and the next. More than post–mortem photography, moving picture technology seemed to provide a portal to an other–worldly experience beyond time measured by the decay of the body. Less morbidly, interaction between people, places and times regarded as separate according to the laws of science became a dominant theme for the so–called 'living' picture shows. Phantom rides, actualities and trick films reiterated many of the visual effects already achieved in magic lantern shows, as they also satisfied a long–standing popular fascination for practices that involved remote interaction in a context that showed a marked ambivalence to the hegemonic claims of science.

Although the Cinématographe appears to have exploited the fruits of the systematic analysis of motion by Marey, Londe, Demeny, Jansen et. al., the public were uncertain about the nature of movement as divisible and these machines appeared to argue both ways by showing the sleight of hand that some procedures of scientific analysis could perpetrate on human senses. Bergson was the respected voice of opposition to positivism and a finite divisible reality. He not only wielded significant populist support for his view of time as the only reality but also gave weight to intuition, that is, the self–conscious positioning of the subject within the object as the only way to break the habitual materialisation of time as space. The extent to which this redressed the argument between determinism and voluntarism can be seen in the public enthusiasm for the paranormal and the transdimensional, even though it was often obvious that hoaxes were being pulled. If, as increasingly science suggested, there were a determining logic to the scientific universe that made our human direction (if not our individual destiny) inevitable, then the strongest alternative appeared to come from inside science itself which had appeared to take a 'wrong turning' in its dealings with telepathy.

Telepathy

Telepathy's great achievement was to provide a rationale that situated scientific objectivity inside the blender. Cartesian method, which had come to dominate classical science, insisted on the separation of the subject and the object, to the extent that, paradoxically the greater the distance between the observer and the observed, the more substantial the truth claim. In contrast, however, telepathy regarded the human mind as a transformative force that could change the world about it. This world was understood as a chaotic field of heterogeneous forces comprising rays, microwaves, electromagnetic emissions and, of course, the force of mind that some gifted or impaired individuals could sometimes under special conditions perceive. When some element in this spectrum of forces found a conduit, a moment of temporary tranquillity was established in the disturbed field that made it intelligible. The form of the conduit could be a copper wire, a gas flame or a particular human mind. Whatever it was, however, the pattern of forces in the world was transformed. As a consequence, the separation between mind and body was a dubious act of self subjugation to an unachievable ideal, while telepathy showed the world to be a product of mind which became clearer the more fully the observer was immersed in it. In this way it was the natural accomplice of the highest achievements of those scientists experimenting with magnetism, electricity, Crooke's tubes and uranium. Telepathy, then, was not just a diversion for the bereaved, the easily ridiculed or the weak minded and foolish but a radical intervention in a single-minded containment of knowledge that reduced the subject to an object. It is little wonder that the emerging professionalism of scientific naturalism was as relentless in its exposure of spiritualistic heresies as it was of new scientific knowledge.

Politicising the predictions of tea leaf gazers, palmists and muscle readers may seem far-fetched at this distance but the interventions of a constituency of the artistic avant garde to recover the epistemological power of the interactivity between mind and world throughout the 20th century can either be seen from outside the blender as a product of history or from inside the blender as a refusal to be determined by the rigid subjectivity of scientific naturalism and Renaissance vision. The preoccupations of some artists with extra dimensions, Theosophy, consciousness or Extensionism has been marginalised by a dubious history of modern art that renders more recent excursions into amplified mind and vision in so-called interactive art as mere technological opportunism, which is wrong and an unacceptable disavowal of the essential nomadism of what that knowledge demands. Sieving chick embryos and macerating sponges to watch them reconstitute may strike the distant observer as a remarkable and even awe-inspiring feat of sub-cellular intelligence, an idea that at the very least shocks to the core our own sense of the monopoly of knowing. Viewed from inside the blender it is all very different or at least provides a very different perspective from which to consider interactivity. For the chick or sponge what to do is neither marvellous nor clever. The individual cells, separated and cast to all four corners of the blender, know where they are because they are of the world and in the world simultaneously, both subject and object, fully conscious of themselves within themselves in ways that perhaps for humans is no longer possible but which artists and psychics can aspire to in spite of the politics of knowledge.

Bibliography

Bottone, S.R. (1893). *Electricity and Magnetism.* London: Wittaker and Co.

Bottone, S.R. (1898). *Radiography.* London: Wittaker and Co.

Brunvand, J.H. (1983). *The Vanishing Hitchhiker.* London: Picador.

Luckhurst, R. (2002). *The Invention of Telepathy.* Oxford: Oxford University Press.

1.5 Bio–electromagnetism: Discrete Interpretations

Nina Czegledy

The earliest accounts of magnetic therapy have been traced to Africa, where ground bloodstone (natural magnetite) has been used for 100,000 years in food and potions. Most ancient civilizations including the Hebrews, Arabs, Indians, Egyptians and Chinese used magnets for healing, as chronicled in Chinese chi's, Hindu chakras and other ancient texts. Aristotle (384–322 BCE) recognized and recorded the therapeutic properties of magnets and the Persian physician Ali Ibn Al–Abbas Al Majusi prescribed magnetic therapy to reduce muscle spasms and gout in 1000 AD (Hacmac 1991). Despite this long–standing and fascinating history, scientific evidence for biomagnetism was first proposed only in the 17th century by J.B. Helmont, who recognized that magnetism plays a significant role in the behaviour of organisms (de Quincey 1993). It was only in the Enlightenment, however, that scientists, by performing repeated experiments, concluded that certain organic tissues contained magnetic properties generating electricity. Electromagnetism was first discovered by the English physicist Michael Faraday (Ramey 1998). Subsequently, James Maxwell clarified the interrelationship between electricity and magnetism in 1864 (Harman 1998). Today the intertwining of these disciplines is clearly recognized; yet up till the 19th century they were considered different phenomena.

The earliest depictions of bio–electricity, notably the discharge of the torpedo mamorata fish, can still be seen on the walls of certain Egyptian tombs dating to 2750 BC (Stillings 1973). Kellaway (1946) traced the ancient clinical applications of the electric fish as follows: 'Headache, even if it is chronic and unbearable, is taken away and remedied forever by a live torpedo placed on the spot. The remedy should be removed lest the ability to feel be taken from the part.' Cladius Galen (AD 130–200), one of the most influential Greek physicians of his time, also experimented with sea torpedoes, concluding, 'The torpedo should be applied alive to the person who has the headache'(Gourevitch 1998).

A new paradigm of bio–electricity was established by the anatomists Luigi Galvani (1737–1798) and Alessandro Volta (1745–1827), among the first scientists to experiment with frogs, who confirmed that frog's muscles generate electricity by themselves. In Paris, the controversial procedures in Franz Anton Mesmer's (1735–1815) magnetic healing salon drew serious criticism from the medical establishment, leading to a temporary decline in magnetic investigations. Over the next couple centuries magnetic therapy was often viewed as quackery and to this day certain treatments –albeit popular –remain contested (Macklis 1993). Nevertheless, scientific research continued and by the turn of the 20th century when Nicola Tesla (1856–1943) pioneered his electromagnetic theories, the biological pertinence of electromagnetism resurfaced. Back in 1890 Tesla suggested the deep–heating value of high–frequency currents on the body and proposed the investigation of electrical anesthesia. Several of his theories are becoming part of current biofeedback techniques (Cheney 2001).

Current bio–scientific interest in electromagnetism was pioneered by Robert Becker's breakthrough experimental work in the 1950s on the bioelectric basis of limb regeneration (Becker 1985). Using salamanders, he was able to show that changes in states of consciousness could be correlated with changes in measurable potentials. Following Tesla's theories of earth resonance, Becker suggested that, owing to our close relationship with the earth's magnetic field, certain Hertz frequencies must have considerable biological significance and that electromagnetic fields can have direct effect at micro cellular levels. 'What is happening today in the field of bio–electricity', said Becker in an interview with Michael Hutchison, 'constitutes a revolution, which is providing us with a greatly expanded vision of the complexity and capabilities of living things.' (Hutchison 1991). In 1963 Baule and McFee first detected a biomagnetic signal, the magnetocardiogram. In the 1980s the emergence of Magnetic Resonance Imaging (MRI) and the subsequent rediscovery of the field of electromedicine revolutionized medical applications of electromagnetism. Since MRI scanners are measured in Tesla units, Nicola Tesla's name is again in daily use and his pioneering theories are re–examined from fresh perspectives.

Over last 50 years, with the renewed interest in bio–electricity, it has been shown that the human body functions like an electromagnetic machine, each body cell containing electrical currents, unique to the individual. It has been confirmed that physical and mental body functions are controlled by electromagnetic impulses from the brain, the pineal gland and the central nervous system. The finding of biogenic magnetite in human tissues was validated when high–resolution transmission electron microscopy and electron diffraction technologies became available in the 1990s (Kirschwink, 1992). Among clinicians, Kyoichi Nakagawa proposed that 'the continuing degrading of Earth's magnetic field, combined with man's electronic environment, is responsible for a broad range of ailments including Magnetic Deficiency Syndrome' (Nakagawa 1976). As a result of these research and clinical studies, it became clear that the balance between negative and positive magnetic forces greatly impacts our physical existence.

By working with various aspects of the invisible and inaudible electromagnetic energies, contemporary artists explore the relationships between transcendental forces and human existence. From Warhol's Electric Chair screen prints to Takis's early Telesculpture to Jasper Johns's brass light bulb sculptures to McClave and Millward's stereoscopic film of the Auroras, the list of artists who have been exploring the realm of electromagnetism is extensive and outside the scope of this text. In keeping with the theme of biomagnetism, I will present projects related to the electromagnetic context of our environment by three contemporary artists. Catherine Richards' *Curiosity Cabinet* examines the blurred boundaries between our bodies and the wired environment. Joyce Hinterding's *Aeriology*, investigates invisible currents. *Bodies of Light,* based on Kirlian aura photography by Marie–Jeanne Musiol, captures hidden 'auras' of objects. The artists utilize in these projects simultaneous readings of scientific information with an aim to question or to better understand our natural and man made habitat as a living medium. Thus these artworks often restore the magic of our bio–environment. On examination, a remarkable intersection is revealed between aesthetics and indiscernible magnetic energies.

In the closed circuit of Catherine Richards's *Curiosity Cabinet* the participant/visitor is supposedly shielded from magnetic interference and becomes 'unplugged' from the 'plugged–in' state of our contemporary surroundings. The visitor climbs inside the cabinet. By closing the doors a closed circuit is created. The cabinet becomes a secure space, an impervious skin from electronic and magnetic currents. Radio signals, microwave and TV frequencies prefer to travel through the metal of the box rather than through the bodies inside. The current seeks to return to its source, thus a thick copper wire provides a pathway for the energy to the earth and grounds the box.

The *Curiosity Cabinet* 'asks what a separated self could mean at a time when such a self is under siege from our own media environments –commented Richards. It holds up the autonomous complete self as a kind of endangered species, a rare collectable on display for a moment. Wealthy Europeans exploring the new worlds often displayed curiosity collections. The media world constantly bathes us in human generated electrical magnetic spectrum. We are permeable to this invisible spectrum as well as its technological displays. To separate ourselves from our media environment it is us, rather than the world, in the cabinet' (Czegledy 1998).

In contrast to this installation, Richards's *Charged Hearts* invites visitors to become part of the electromagnetically charged atmosphere. The work consists of a physical installation and a virtual on–line component. By walking on the glass and steel platform and touching one of the bell jars containing a blown glass heart, a sensor is activated and the glass object transmits a phosphorescent glow. *Charged Hearts* also includes a web game, which takes the slippery language of electromagnetism and shows it mirroring the language of emotions. Simultaneously the spectator's heart rate data is sent to the web page on the Internet. Thus the *Charged Heart* and the *Curiosity Cabinet* installations complement each other —one by plugging the visitor into the electromagnetic atmosphere, the other by unplugging him/her (Porter 2000). Evoking and working with these metaphors Richards, in our visually privileged world, is directly accessing our other senses.

From her early work with acoustic objects that were designed to resonate, Hinterding proceeded to auditory bi–products of our electromagnetic environment. Hinterding's *Aeriology* has been described as a project for the unfolding of the ethereal as a 'machine for techne of the invisible'. The installation contains 20 kilometres of wrapped wire to form an energy gatherer. The coils reveal through sympathetic amplification the activity of the unseen. The 'line' of the wire literally gathers and reconfigures energies, turning refuse static to potential by a process of realignment of the subtle bodies of particles (Hinterding 1995). Joyce Hinterding's interest in electromagnetism began with the exploration of auditory facilities. She became fascinated by sound, by the phenomenon of sympathetic resonance, 'the ability for one vibrating body to activate another. It is an interest in 'this that exists between things', rather than 'things' that characterized my initial approach to sound. A concern with the dynamic nature of the world, the relationships between things'. Thus commented Hinterding in an interview with Josephine Bosma. Later, this interest in acoustic amplification developed into an interest in electronics, firstly with electronic amplification, then with electronic sound production, synthesis, feedback and eventually the auditory bi–products of our electromagnetic environment. Sound is a physical expression of vibration. However, when working with the electromagnetic, one is working with dematerialized activity. To understand and develop work concerned with massless activity Hinterding built a VLF (very low frequency) antenna. The sounds in this section of the radio spectrum are a product of natural atmospheric electrical activity. 'The VLF range is only a section of the ambient electromagnetic environment we live in and I use the term electromagnetic landscape to look at the local transmission environments' (Bosma 1998).

Bio–electrography, or Kirlian photography, consists of placing an object or body part directly onto a piece of photographic paper and then passing a high voltage across the object. The Kirlians proposed that the photographic paper will become exposed and will show a glowing 'aura' around the object. The 'aura' is supposed to be a product of 'bioenergy' or 'bioplasma'. Kirlian photographers claim that different moods and levels of psychic power will show up in these photos. Konstantin Korotkov, a Russian biophysicist exploring stimulated electrophotonic emissions around the human body and other objects, developed the new Gas Discharge Visualization technique (Korotkov 1998). Marie–Jeanne Musiol, while investigating metaphors conveyed by photography, began to use electrophotography to capture the invisible energies surrounding objects, a fundamental although unseen component of life. Through a series of electrophotos on the same subject, the artist revealed various phases of the aruora encircling objects such as plants or a human finger. By inducing an electromagnetic field around an object, photographing and videotaping the luminous emanations, she claims that she has conveyed meaningful information on the physical and psychic state of objects and their interaction with the environment. Musiol's elegant images reveal the complex, alluring mystery of living organisms from yet another viewpoint. Her book containing text and images on Kirlian research, as well as a scientific description by Prof. Konstantin Korotkov from Russia, has been recently published (Musiol 2001).

In summary, despite extensive bioelectromagnetic research, many essential questions remain unanswered. Is the transmission of psychotronic thought projection in a subject's mind via low frequency waves the same phenomena that occurs with telepathy or telekinesis? What is the final verdict on the impact of very low frequencies on the human body? Do all of the phenomena derive from the same elemental force, an unseen energy that can manifest within various frequencies and spectrums? Are the currents of energy, which physicians utilize and artists explore, the very same electromagnetic, biomagnetic or psychotronic energies that yogis use to modulate their brain wave frequencies during meditation? Recent theories contest the chemical–mechanistic model of biological systems, which have dominated the field of medical science. Paradigms are shifting to alternative interpretations based on complex systems, non–linear dynamics and chaos theory. This new vision of phenomena seems to restore electromagnetic energy to a position of prominence. By revealing aspects of interactions between the electromagnetic field and biological systems the investigations by scientists unveil mysteries such as undefined diseases or altered states of consciousness. The art projects seem to recapture the often–lost connection between science, art, technology and the natural world. In conclusion, I would like to quote from Robert Becker: 'What is emerging is a new paradigm of life, energy and medicine' (Hutchison 1991).

Bibliography

Becker, R.O. and Selden G. (1985). *The Body Electric: Electromagnetism and the Foundation of Life*. New York: William Morrow & Company, Inc.

Bosma, J. (2000). *Sound Art: Joyce Hinterding*.

http://amsterdam.nettime.org/Lists–Archives/nettime–1–9808/msg00074.html

Cheney, Margaret. (2001). Tesla –Man Out of Time. New York: Touchstone.

Czegledy, N. (1998). *Aurora: An exhibition of installations*. Toronto: InterAccess Electronic Media Arts Centre, exhibition catalogue. p. 9.

de Quincey, Christian. (1993) 'Bioelectromagnetics: Old Roots of a New Science'. *Noetic Sciences Review*, vol. 28, pp. 30–2.

Gourevitch, D. (1998). 'The Paths of Knowledge: Medicine in the Roman World', in M.D. Grmek (ed.). *Medical Thought from Antiquity to the Middle Ages*. London: Harvard University Press.

Hacmac, E.A. (1991). *An Overview of Biomagnetic Therapeutics*.

Harman, P. (1998). *The Natural History of James Maxwell*. Cambridge: Cambridge University Press.

Hinterding, Joyce. (1995). *Aeriology*.

http://hosted.at.imago.com.au/luminoska/aer.htm

Hutchison, M. (1991). *High Voltage: The Megabrain Bioelectric Interviews*. Megabrain Report, vol. 1.

Kellaway, P. (1946). 'The part played by electric fish in the early history of bio–electricity and electrotherapy'. *Bulletin of Historical Medicine*, vol. 20.

Kirschvink, J.L. et al. (1992). 'Magnetite biomineralization in the human brain'. *Proceedings of the National Academy of Sciences*, 89 (1992):7683–7687.

Korotkov, K. (1998). *Aura and Consciousness –New stage of scientific understanding*. St. Petersburg: Federal Tech University SPIFMO, Kultura.

Macklis, R. (1993). 'Magnetic Healing, Quackery and the Debate about the Health Effects of Electromagnetic Fields'. *Annals of Medicine* 118(5), pp. 376–83,

Musiol Marie–Jeanne. (2001). *2001 Corps de lumière/Bodies of Light*. Montreal: Axe Néo–7 art contemporain.

Nakagawa, Kyoichi. (1976). 'Magnetic Field Deficiency Syndrome and Magnetic Treatment'. *Japan Medical Journal*, no. 2745.

Porter, D. (2000). *About Harnessing Forces*. Halifax: Dalhouse Gallery, Engaging the Virtual. Exhibition catalogue.

Ramey, D. W. (1998). 'Magnetic and Electromagnetic Therapy'. *The Scientific Review of Alternative Medicine*,vol. 1, pp. 1–6.

Stillings, D. (1973). 'Artifact: The piscean origin of medical electricity;. *Medical Instrumentation*, vol. 7,no. 2, pp.163–4.

www.biomagneticsofbeverlyhills.com/history.html

1.6 MEDIATE: Steps Towards a Self–Organising Interface

Paul Newland, Chris Creed and Maestro Ron Geesin

Background to MEDIATE

Preliminary foragings that have now led to MEDIATE (Multi–sensory Environment Design for an Interface between Autistic and Typical Expressiveness) have been reported at the Inaugural Consciousness Reframed Conference in 1997 by the authors (Newland and Geesin 1997). Progress towards gaining funding for the eventual project was tortuous and is worth reviewing here to highlight the continuing difficulty in accessing an appropriate funding body for multidisciplinary research. Our project originally applied for funding to the UK's Engineering and Physical Sciences Research Council (EPSRC) –their conclusion was that it was not science, no hypotheses. They suggested the UK's Arts and Humanities Research Board (AHRB). AHRB decided they could not consider it as it was therapy and suggested the UK's Medical Research Council (MRC). We knew it was not therapy, so instead we sought funding from the EC 5th Framework, IST Programme,1 which was genuinely funding multidisciplinary research and not just paying lip service to the concept.

Our team2 draws members from the following institutions, company and organization: the University of Portsmouth (overall coordination, vibrotactile); the Hogeschool voor de Kunsten Utrecht (HKU), The Netherlands (sound, participant's interacting pattern detection system); the Pompeu Fabra University (UPF), Spain (visuals, Intranet, and DVD record); Institute of Psychiatry, Kings College London (psychological evaluation); Show Connections Ltd. (technical design, environment construction); Autisme Europe, a European–wide parents' organisation.

So far we have had opportunity for four full partners' meetings. They have all been extremely fruitful and essential to gaining mutual understanding in what is not only a multilingual mind space but one with multidisciplinary vocabularies. Opportunities for face–to–face communication and joint hands–on resolution of technology connection should always be the top budget priority, an obvious insight that our experience can most certainly endorse.

MEDIATE's purpose

The spectrum of competencies and international field leaders within our team was a significant factor for enabling success in gaining EC funding but was primarily necessitated by the ambition of MEDIATE. Our purpose in creating the

MEDIATE Project was to design, produce, build and validate an 'intelligent', immersive, multi-sensory, interactive environment that could react to each unique Person on the Autistic Spectrum (PAS), allowing expression of multi-sensory experiences: creations which could be replayed and communicated to others. A further criterion was that this environment should be transportable.

MEDIATE cannot be neatly categorized as art, science, training or therapy. However, it is intended that MEDIATE will lead to new expression (which may be 'artistic') and which should be subject to 'scientific analysis' which will lead to new understandings (which may be applied to 'training'). Essentially its aim is to give fun and allow expression and for the user to feel in control (perceiving a sense of agency). As such it may well become 'therapeutic'.

The various component technologies of MEDIATE[3] are being designed to allow interaction individually in visual, auditory and vibrotactile modes that can increase in complexity within a modality and through cross–modality complementarity and counterpoint. Such complexity is prompted by and under the continuous control of the autistic child by the 'brain' of the environment reacting to pattern detection of the individual –the individual's interactions creating a unique signature. This is our goal, though at present we have yet to reach the fine–tuning phase.

Current Status

At present MEDIATE is eight months from the end of its EC funding framework. The full environment is currently being assembled in Boathouse No.4, Historic Dockyard, Portsmouth. Here the following components are being integrated.

An irregular walled hexagon is held in place by a Tri–lite rectilinear support frame attached to a raised floor formed from 1m2 panels, roofed with a grid of infra–red lighting –the overall physical footprint is 7.5m2 with a 3.5 m height.

2) Walls 3 and 5 are full coverage back projection screens, whilst walls 2 and 4 carry a touch sensitive, multi–textured branched bas–relief and vibro–equipped floor to ceiling curvi–linear impression tubes respectively. Wall 6 is left plain with a disguised viewing grill (through which parents/carers and psychologists can observe) and wall 1 holds the entrance which is reached by an exterior ramp, semi–shielded by a fabric covered arched passageway.

3) The area behind wall 1 houses a technician and the computers operationalising the two back projectors, image creation, EyesWeb[4] user position sensing analysis and camera feeds; microphone and pressure sensing capture with eight speaker sonic and multi–vibration feedback units controlled by MAX/MSP interaction software; messaging protocols, decision–maker and signature–pattern detecting software.

4) Each 1m² floor panel is designed to sense movement and enhance the sound properties of footsteps by means of a sandwich construction incorporating 'crunch' fabric and piezo–electric pick–ups. The floor expanse is occupied, off–centre towards wall 6, by a 0.8m radius mushroom that is touch sensitive and vibro–responsive.

Once full assembly is complete the environment will be fine tuned until September when it will then be taken to locations in London, Hilversum and Barcelona for evaluation trials with autistic children and their families coordinated by the Institute of Psychiatry, King's College.

Potential Developments

Throughout the creation of MEDIATE's range of modality interfaces and their intended coordinated interaction we have endeavoured to adhere to four design principles.

- to encourage the individual towards systemic involvement and away from component fascination

- to entice a fully embodied awareness, so the individual discovers they are within a body which has boundaries and this embodied self has agency over the environment's response

- to reassure whilst 'opening the door' to the playful, the environment having a capacity to build a history of the individual's tolerance for the unexpected

- the interface components should attempt to respond to the unique actions presented by the individual and avoid demanding decryption as a means to pull interaction

Of course, our prime concern in the present manifestation of MEDIATE is that PAS can truly gain an experience of agency, that their action pulls congruent reaction without them feeling they have merely stumbled across some pre–determined interaction sequence. Current thought from the perspective of integration is to evolve a user's interaction with one modality into a level of interaction that gains increasing call on the other two available modalities. Such an increasing accompaniment and weaving together of sonic, image and vibrotactile interface dimensions must clearly be under the agency of the individual who is inhabiting MEDIATE.

To facilitate this gradual, user–elicited integration there would be gradients of responsiveness for each sensory component. For example, rather than first steps into the MEDIATE environment generating a response on sonic, image and vibrotactile sensory spectrums all at once, that relative direction of movement and position will instead initially favour one sensory spectrum in particular. This necessity for individuals to be allowed to invoke their own interaction pathway is asking for a capacity to essentially self–construct an interface in real time. For the user–environment dyad to have this capacity, the environment at some level needs to have a 'brain'.

Part of the team at HKU has concentrated on developing pattern detection protocols –signature software. Originally conceived to distinguish the different playing styles of jazz musicians, the more generic ability to extract patterns from the sensing of human action has been considerably refined and implemented for analysing the data streams from MEDIATE's multi–modal sensing systems. Theoretically within the limits of time expanse and sensing resolution any behaviour sequences recurring more than once will be flagged as a pattern of interaction. For the purposes of the environment's engagement with the PAS, repetitive or non–generative behaviour can be countered by a look–up table of sensory modality feedback and overall response options, e.g. dimming, shifting modality priorities and/or sensitivities. However, novelty, the expression of diversification in engagement, can be encouraged both within and across modalities through amplification, echo and counterpoint.

Within the parameters of the possible modalities' feedback, the user–environment dyad can then self–organise towards various states of interface coordination. How far these states are catered for by the decision–making protocols of the 'brain' and to what extent they can be unexpected and arise spontaneously will become evident as we fine tune.

Bibliography

Newland, P.M. and Geesin, R. (1997). 'A Web of Senses in an Autistic Universe', in Scott, R. (ed.), *Proceedings of the First International CAiiA Research Conference on Consciousness Reframed*. Newport, Wales: University of Wales.

Notes

1. The MEDIATE Project (IST–2000–26307) gratefully acknowledges the financial support of The European Commission, Information Society Technologies, 5th Framework Funding Programme.
2. Team members and contacts: http://www.port.ac.uk/mediate/projpartners.htm
3. For full details refer to http://www.port.ac.uk/mediate/medipuba/home.htm
4. UPF are collaborating with Camurri's EyesWeb team (http://www.eyesweb.org/) to extend some analysis protocols tailored specifically for MEDIATE.

1.7 Facts about 'P–E–M' (Psycho–enhanced Memberships) you must know

Armando Montilla

The psychological disposition for the aesthetic perception of urban space can be understood as dissociation, a gap in our memories and hallucination. The disparate dispersed and transitory elements inherent in the modern city no longer come together –they can't be understood as a unity. There is no inner image of the world, it is no longer possible to produce a synthesizing idea to find a correspondence between the inside and the outside. These excessive aspects of contemporary life allow ourselves to create provisional identities.

(Bittner 2001)

The study of the brain is not just a biological science. It is also chemical, electrical, cultural and psychological. In fact, ever since the Hixon Symposium at the California Institute of Technology in September 1948,[1] cognitive science has been regarded as an interdisciplinary field. Most people date the beginnings of cognitive science from this symposium, the title of which was 'Cerebral Mechanisms of Behaviour'. Presenters included John von Neumann from mathematics and Karl Lashley from psychology.[2] Today, six major disciplines work together under the banner of cognitive science to explore the workings of the brain/mind: psychology, neurobiology, information science, anthropology, philosophy and linguistics. Some academic programmes emphasise information science, while others emphasise neurobiology. But the field is genuinely interdisciplinary.

Traditionally, the brain has been defined as the subject of biology, while the mind belonged to psychology. That dichotomy became entrenched when Descartes supposedly made a deal with the church, that they would leave him alone if he professed only to study the brain and left the mind to the theologians. Current thinking does not view these two subjects as separate. One makes no sense without the other. Without the brain, there is no mind; with no mind, there is no brain. The essence of this viewpoint is revealed in the study of *psychoneuroimmunology* (or *PNI*). *PNI* has identified the capacity of effortful thought to affect the levels of immune agents in the body, as well as the capacity of immune agent levels to affect the quality of thought and mood. To paraphrase Dr William Osler (Princeton Weekly Bulletin 1999), it is often not so important what kind of disease the patient has but what kind of patient has the disease.

Recent findings in Cognitive Psychology have shown that thousands of freshly born neurons arrive each day in the cerebral cortex, the outer rind of the forebrain where higher intellectual functions and personality are centred (Montilla *IDENTI*–Park: The Cross–dressing of the City*). The forebrain is knowingly mainly responsible for memory and intelligence and it includes the cerebrum and its left and right hemispheres connected to each other by the corpus callosum and covered by the grey cerebral cortex. Ever since German psychiatrist Hans Berger investigated the physical basis for mental functions and devised the procedure of electroencephalography (EEG), which became a standard diagnostic tool in neurology by way of recording human brain electrical waves, electrical activity seems to be the key for personality and memory.

Quotes from IDENTI*–Park, novel (2002)

...IDENTI*–Park was created by the Holding Company implemented by the various multinational corporations that took possession of the old Europaviertel area [in Eurofrankfurt]. The purpose was to put in place a Theme Park, based on the premises of a parallel identity enjoyed by selected members, who had exclusive access to this area of the city. The chosen identity would be implemented and given to members in a sponsored manner, by one of the multinational brands participating in the newly–created Urban Park...

(Montilla, IDENTI*–Park, p. 13)

...The new interactive IDENTI*–Park psyche–enhanced membership ... uses the latest NSTT technology ... But it also activates a sequence in the cells at the part of the brain, which controls our personality and abilities ... It is a breakthrough technique, which identifies secret desired roles or unfulfilled fantasies within the subconscious, and encrypts them in that part of the brain ... When our future member manipulates with his or her own thoughts the different options that are being presented, he or she does not suspect he or she will actually become that person in a matter of a couple of hours ... without remembering anything from his or her previous life ... In this way the acting is most realistic, rather authentic...

(Montilla, IDENTI*–Park, The Free–Cities, p. 45)

The 18 facts about IDP–PEM (IDENTI*–Park Psycho–enhanced membership) you must know

Fact 1: What exactly is an IDP–PEM? (IDENTI*–Park Psycho–enhanced Membership)

Available to the all IDENTI*–Park Members since 2032, IDP–PEMs (IDENTI*–Park Psycho–enhanced Memberships) are individually awarded and neuro–electronically induced alternative IPB (Info–projected Beings), which have pre–determined temporary existence. Once an IDP–PEM is activated, it replaces the standard Neuro–psychological Being for a specific period of time.

Fact 2: What is an IPB? (Info–projected Being)
An IPB (Info–Projected Being) is a trademark of the IDENTI*–Park TRANS–ROLE Technologies Division. IPBs are the result of a projected alternative neuro–psycho character based in closest–probability and resulting from the information gathered from a data–produced MPP (Memory–Personality Prototype) chart.

Fact 3: What is a MPP? (Memory–Personality Prototype)
A Memory–Personality Prototype (MPP) is the ensemble of elements that neuro–psychologically defines an individual, as copyright–registered by the Neuro–psychology Division of ACRE Inc. (Advanced Cognitive Research for the Entertainment Industries Inc., a multiversal company) ...

Fact 4: How is an IPB tm created?
Through latest state–of–the–art NSTT third generation technology, an electronically made scan of the forebrain calculates a history of chemical and electronic activity in the neurons of this area, reconstructing non–identified long–term memory groups. The processed info produces an alternative MPP (Memory–Personality prototype).

Fact 5: How is an IPB implemented?
Once the alternative MPP (Memory–Personality Prototype) is completed, it conforms the IPB (Info–projected Being), which is then de–codified as receivable data and transmitted by low–band NSTT sequences to the neurons at the frontal grey brain cortex, where personality and abilities groups are located. Low–band NSTT sent at this area create a medium–range level of electrical activity intensity at the neurons' dendrites –thus implanting new abilities and personality groups –while high–band NSTT sequences sent to the mid–grey brain cortex overlap existing memory groups with newly–implanted ones based on the projected MPP. In this way the IPB (Info–projected Being) replaces the standard Neuro–psychological Being.

Fact 6: How long does an IPB last?
First IDP–PEMs (IDENTI*–Park Psycho–enhanced Memberships) offered to IDENTI*–Park Members in 2032 included IPBs (Info–projected Beings) that lasted approximately 11 to 13 months. Today, thanks to advanced development on NSTT third generation technology, IDENTI*–Park TRANS–ROLE Technology Division is able to produce IPBs lasting to exact pre–determined periods of time, thus giving a more secure and predictable implementation.

Fact 7: Why can IPBs only be temporary and how are they set to last?
Electrical and chemical activity produced at the front and mid-grey cortex brain area neuron's dendrites (by medium and high band NSTT transmission) is pre-programmed to have a pre-determined life cycle according to a codified sequence imprinted in the transmission. Once the life cycle is over, standard electrical and chemical activity takes place, allowing for the standard Neuro-psychological Being to manifest again. Naturally, every individual's PNI (psychoneuroimmunology) varies, which requires IPB charts to calculate intensity and allowable PNI in every person.

Fact 8: Do I remember anything from my normal life under IDP-PEM?
Electrical and chemical activity linked to the IPB (Info-projected Being) act at higher amplitude than those linked with the standard Neuro-psychological Being. In this manner our standard existing memory and personality groups become inactive (what used to be commonly known in old non-cognitive psychology as 'amnesia'). While this happens the memory and personality groups implanted in the IPB (Info-projected Being) are active until their life cycle ends. The result is that you will not remember anything, neither will you act according to your previous standard Neuro-psychological Being.

Fact 9: Do I remember anything from my IDP-PEM after it is over?
As memory groups are created in an cumulative manner (through a visual-spatial and phonogical loop, involving both hemispheres of the forebrain), once an IPB (Info-projected Being) life cycle has come to an end, all personality and ability groups linked to it will vanish, while memory groups will not. So you will come to remember everything you lived through, did and experienced during the IPB life cycle period.

Fact 10: Is an IPB a product of my unsatisfied desires/fantasies?
It apparently is but, in reality, NSTT scanning of the electrical and chemical activity history (of the neuron's dendrites at the front-grey cortex brain area) can only 'reconstruct' not 'guess' repressed memories and/or desires. The reconstruction occurs by a projected memory and sensations group matrix, elaborated based on an optimised model of electrical and chemical ratio-activity detected. By laws of probability, the new 'desires', 'emotions' and 'personality' linked to the IPB (Info-projected Being) will correspond with the 'real' ones with a minimal error margin.

Fact 11: Why should an IDP-PEM be based on an IPB only?
IDENTI*-Park extensive studies of standard non-psycho enhanced Memberships (previously offered in IDENTI*-Park City-resorts) concluded that Members were unable to authentically 'act' their new MPP (Memory-Personality Prototypes), owing to interference caused by memory and personality groups belonging to their

daily standard one. By temporarily 'suppressing' the standard Neuro–psychological Being and replacing it with an IPB (Info–projected Being) the new MPP becomes an extremely authentic and accurate one.

Fact 12: What's the use of not knowing when I am under an IDP–PEM?
While this seems paradoxical, not knowing your 'real' and previous MPP (Memory–Personality Prototype), while your IPB (Info–projected Being) is implemented assures to Members a more satisfying recollection of their 'someone else' experience, without any risk of contamination produced by their previous MPP. In most cases the remembered revelations and recollections are of an extraordinary nature.

Fact 13: Will my memories/personality/abilities suffer any damage after my IDP–PEM is over?
Conclusion of an IPB's (Info–projected Being) life cycle, together with the cessation of electrical and chemical activity linked to the IPB (Info–projected Being), guarantees full restitution of the original MPP (Memory–Personality Prototype) by bringing the standard Neuro–psychological Being to the surface intact, as of the last moment before the new IPB (Info–projected Being) was implemented. As a measure of caution, it is recommended that you purchase a NPBMP (Neuro–psychological Being Memory–Print) for an extra fee.

Fact 14: What is a NPBMP (Neuro–psychological Being Memory–Print)?
Each time IDENTI*–Park elaborates an IPB (Info–projected Being) for a new member a NPBMP (Neuro–psychological Memory–Print) is produced, as a footprint of your standard MPP (Memory–Personality Prototype tm). In case of failure, NSTT transmission can be used to implement it, bringing back a clone MPP of the original 'you' as it was registered by your NPBMP. As of today, NPBMPs have been elaborated only as a precautionary device and their implementation has been extremely rare.

Fact 15: Is a NPBMP guaranteed to last as long as my original MPP (Memory–Personality Prototype)?
As its implementation has been very rare, NPBMP's (Neuro–psychological Being Memory–Print) performance has only been measured through simulation–projected cases. In the only one monitored for real implementation, the individual under study was given an extended life cycle new MPP, which showed no alteration/difference from his previous MPP as it had been recorded (though there were slight changes in the appetite for certain types of food, which were not present before). This case is presently under study and is being monitored. In another case where NPBMP implementation was recommended, after the IPB (Info–projected Being) life cycle finished, bringing the IDP–PEM (IDENTI*–Park Psycho–enhanced Memberships) to an end, the individual in question refused to

have NPBMP administered and chose to continue experiencing a continuously administered IPB (Info–projected Being). Simulated projected cases to date show that NPBMPs need revision and/or re–administration every five years or so.

Fact 16: What is the longest time an IDP–PEM can be implemented for?
Typical IDP–PEMs (IDENTI*–Park Psycho–enhanced Memberships) are limited to last a maximum of approximately 16 months. However, there seems no limit on the number of times an IDP–PEM can be implemented to the same Member. There are recorded cases of IDENTI*–Park Psycho–enhanced Members who have continuously been under their IDP–PEM (through continuous requested re–implementation) for more than three years to date, without any recorded problems.

Fact 17: Can the original MPP tm resurface while under IDP–PEM?
There have been a few recorded cases when traces of the original MPP (Memory–Personality Prototype), linked to the standard Neuro–psychological Being, has slightly resurfaced during IDP–PEM tm active period. In these cases, according to previously agreed clauses between IDENTI*–Park tm and IDENTI*–Park tm Psycho–enhanced Members, IDP–PEM tm re–implementation has been performed. Continuous and intensively close monitoring of every active IDENTI*–Park Psycho–enhanced Memberships tm is conducted, in order to immediately detect any traces of original MPP (Memory–Personality Prototype tm) that might resurface, causing damage to the characteristics of the IPB (Info–projected Being tm) and protecting the integrity of the IDP–PEM tm.

Fact 18: Can an IDP–PEM tm be interrupted once is implemented?
IPB's (Info–projected Being tm) life cycles are programmed to last according to IDENTI*–Park tm Psycho–enhanced Members' specifications at the time of signing the IDENTI*–Park Psycho–enhanced Membership tm Contract. An IDP–PEM tm can only be interrupted before its life cycle's end by NPBMP's (Neuro–psychological Being Memory–Print tm) implementation. This would only be performed in case of IPB's (Info–projected Being tm) malfunction becoming hazardous for the IDP–PE–Member (Montilla, *IDENTI*–Park, Part III: Corralito, The Liberated City*, pp. 102–8).

The facts mentioned above have all been reflected in fictional writing. I do believe that the line separating cognitive science and science fiction does become more blurred each day, so certain scenarios can only be envisioned via literature. *IDENTI*–Park* is a fictional writing exercise which shows, among other things, how far science might go to provide personal satisfaction.

Bibliography
Bittner, Regina. 'A Visit to IDENTI*–Park', in *Urban Detours*. Bauhaus Kolleg II 'Event City'. Dessau: Bauhaus Dessau Foundation, 2001.

Lashley, K. (1951). 'The problem of serial order in behaviour', in L. A. Jeffress (ed.). *Cerebral mechanisms in behaviour: The Hixon symposium*. New York: John Wiley & Sons. http://peace.saumag.edu/faculty/Kardas/Courses/CS/Student%20Pages/Maeweather/HistPeop.html

Montilla, Armando. 'IDENTI*–Park: The Cross–dressing of the City', in Regina Bittner (ed.). *Die Stadt als Event*. Frankfurt: Campus Verlag, 2002.

Montilla, Armando. *IDENTI*–Park, Part II: Ahbittipura, The Free Cities* (unpublished).

Montilla, Armando. *IDENTI*–Park, Part III: Corralito, The Liberated City* (unpublished).

Princeton Weekly Bulletin. 25 October 1999, vol. 89, no. 7.

Internet and CD resources

On the brain: *Stuff of Brains*:
www.cogs.susx.ac.uk/lab/nlp/gazdar/teach/atc/1999/revman/markb.html

Princeton Weekly Bulletin. 25 October 1999, vol. 89, no. 7: www.princeton.edu/pr/pwb/99/1025/

The Human Brain:http://dryden.lakehadu.ca/~gwwintle/brain/

Cosmic Baseball Association. *1999 Mindland Brains*: www.cosmicbaseball.com/99mbr.html

On *IDENTI*–Park*:

IDENTI–Park: The Cross–dressing of the City* (PDF):
www.bauhaus–dessau.de/K2T3/identi–park

IDENTI–Park, Part II: Ahbittipura, The Free Cities* (review, synopsis and excerpt). 'Urban Drift' Festival: 'From Formalism to Flux'. Berlin, Germany. October 2002:

www.urbandrift.org/projekte/2002/nightspace/pro_ud_night_arm.html

IDENTI–Park, Part II: Ahbittipura, The Free Cities* (excerpts from Reading) 'Urban Drift' Festival: 'From Formalism to Flux'. CD Publication. Berlin: Urban Drift, 2003.

1.8 Happenstances

Demnievska Evgenija

Happenstance is an art event composed of simultaneous actions performed in different environments or in different cities linked by the Internet. The totality is a synchronized real time event that is singular and non–repeatable, realised in good circumstances with a simultaneously generated constructive energy.

As much as we are interested in new technical possibilities, we are interested in results based on the conditioning of the consciousness of the participants before the event. The determinant of the quality of the result of the Happenstance is the exact level of consciousness each participant has about the project, as well as his responsibility and ethical behaviour.

If members of the audience decide to participate they have to be initiated before they enter the performing space (in situs or on the net). The initiation attempts to match the experience of the participants from the audience with the consciousness of those who are well prepared.

Initially the project is an individual creative vision, discussed with a group, before the final formulation of the rules is accepted by consensus.

The events happening simultaneously in different places are in correlation and the connection and relationship of many is based on a resonance.

After numerous experiences, we have concluded that the physical performance directly seen does not really contribute to the quality of events using the new technology. Thus the place of the action in real space is hidden, not directly exposed to the view of the public. It is accessible only to the participants and to those who wish to join from the audience, guided by the 'Co–Resonance' team members.

The Happenstance is to be seen only on screen.

The participants themselves structure the event, achieving a mental attitude whereby they have the confidence to know how to act spontaneously and with freedom, guided by team members who have experience.

The overall intention is to mix different levels of reality, including screen images from the real world, images reflected in mirrors and images produced by the computer (images of the syntheses).

When the camera captures the images in the mirror it often creates an optical illusion erasing the difference between the real images and the images reflected in the mirror, producing twice reflected images, endlessly reflected, escaping the difference. The images produced by computer, videos, 3D, etc. are made before the event.

As to timing, the moment one sends these images to be seen among others is decided spontaneously by means of intuition. These moments can be chosen by the person in charge who might watch the Happenstance or who can send them successively. These images are presented simultaneously with the images captured by camera from the places where the action is happening in the real space. The form, although manifesting itself as a totality, is always segmented, comes from different locations and is visible on several screens. It is a poetical space generated by the people participating from a distance, who cannot see the totality at the moment they act. 'To be' and 'to know' are always separated.

The spectator in this crossover space can either focus on the structure of the project itself or search for meaning in the images on the screens or, even more importantly, in the relationship of these images.

Andre Breton and the surrealists talked about the 'objective hazard', a singular non–repetitive event that can make visible the hidden sense of the event. It is supposed that the event has a meaning.

Carl Jung talked about synchronicity as events that are related to chance. These events make sense to the person to whom they are happening. Usually two or more events happen without a cause and produce an epiphany for the person. The sense is not created on a rational basis and might be interpreted in different ways by different people. The information is given to the one who is concerned by the coincidence and depends on his/her interior, subjective world.

Instead of linear time, synchronicity addresses some other subjective time, where one perceives repetition and cycles. This subjective time, which manifests itself in synchronicities, is an element of the unconscious, of our interior being, and its significance will never have a single interpretation.

The synergy of the participants being in different locations, the flux of the energy turned in such a direction that everything happens better than it was planned, is a basis for pleasant surprises mixed with horror, as a manifestation of the sublime.

The Happenstance is a multi–professional teamwork of artists, technicians, observers, theoreticians and advisers. Each participant has a precise role and is a co–creator of the event. The supervisor steps in with a decision in the eventuality of a crisis, when no time is left for discussion.

The entire project represents an important investigation of energy, imagination and technical competences for a result that is not completely known in advance.

Regarding partnerships for the realisation of the Happenstance, we come to the question of the interpretation and the classification of the project itself.

To paraphrase the sociologist Nathalie Heinich in her book *Le Triple Jeu de l'art contemporain*:

> ... the questions are no longer concerned with the sense and the value of the work, but with its nature. We are passing from one problem of aesthetic qualifications to the problem of cognitive classification between discontinued categories: is it a work of art? If yes, where could it be classified? And if not, what is it? Since the aesthetic appreciation commands the value attributed to the work, it is distinct from the original intention of how the artist wants his work to be interpreted and appreciated which defines its status...

> ...The work of classification consists first to enter the work mentally into the cognitive category, before it can find a place for its realisation or its presentation.

> At the same time the difficulty of classifying the project contributes indirectly to the artist: the effect produced by the cognitive floating reinforces the legitimacy of the dossier of the artist, because the transgression of the usual categories of perception and classification is one of the fundamental competences of contemporary art...

> ...The question of the nature of the work leads to spontaneous interpretations that are attempts to reduce the uncertainty created by the transgression of the cognitive frame.

> The difficulty in framing the work into the forms of existing classifications calls for another treatment where we attempt to achieve rationalisation, with the help of interpretative hypotheses that allow us to formulate opinions. These interpretations reduce the gap between the economic norms of spending and rewards, as the results are spiritual and cognitive not economic...

> ... If confronted, in spite of all presented information, with an unclassifiable phenomenon, the artists improvise the attitude of the relation to the world they have in their disposition: bringing down the phenomena to more familiar registers, either playing with it by making it go through mental transformation of the category or by appropriating the fragments, the traces, as a private gesture, the process of personal appropriation accomplished publicly by the author.

> The function is, like the expression of an opinion, to permit the individual to regain mastery over the project which, being inadequate to the familiar schemas, is menaced.

There is then a need to establish an order by using the 'common' sense that will find consensual criteria that are currently missing...

(Heinich 2002 pp. 190–5)

To return to the Happenstance, events demanding the extreme concentration of participants, whose state of mind should allow almost a telepathic connection alongside the technological, were already formulated in 1995–6, the beginning of the expansion of the Internet, and we invested incredible energy in realising them. We succeeded first in 1997 and then in 1998, getting support mostly from artists, and associations created by artists, and from web–bars as well. We also got support from the City Museum in Skopje and galleries in Sofia, Paris, Krakow, etc., mostly for the realisation of the event and partly for the technology.

Although extremely precious for our experience, this support was not sufficient for the optimal realisation of the project. Neither were these locations useful places to provide the better conditions necessary for the future realisation of the project, as the frameworks for experimental projects using new technology did not exist in these institutions. We were a sort of clandestine project in the framework of the programme of the institution, supported by people who had influence and sympathy for our position but who did not have a budget at their disposal to help with the optimal realisation of the project.

Items that were published in the press were mostly written by journalists. Art critics prefer well–trodden categories where they have more expertise.

So the forum, being far more flexible and innovative than art institutions, appears, on the one hand, to be a new possibility for the artist to gain integration by presenting the project publicly as well as discussing the problem it is approaching. In order to get the work re–evaluated, the artist instigated transgressive acts, to affect, through this participation, the commentaries of the critics. The forum is a place to get more information about new ideas and practices directly from artists and researchers and to better understand where to search for potential new partners for the realisation of the projects they are proposing.

By expressing their position in texts and books and analysing new phenomena, artists are contributing to the reformulation of the framework which can integrate works of other artists dealing with these phenomena. They are becoming an important point of reference indicating the alternative possibilities for changing the cultural and the educational context today.

Bibliography

Heinich, Nathalie. (Janvier 2002). *Triple jeu de l'art contemporain.* Paris: Les Editions de Minuit.'

1.9 Ontological Engineering: Connectivity in the Nanofield

Roy Ascott

The questions at the frontier of knowledge today are precisely: where is mind, what is consciousness? Is consciousness an epiphenomenon of the brain or is materiality an epiphenomenon of consciousness?

In so far as new media arts are concerned, consciousness has been at the top of the research agenda for almost a decade, as exemplified in such conferences as those organised by the Planetary Collegium since 1997. Similarly, since 1994, debate and discussion that might lead to establishing a science of consciousness has been the goal of the University of Arizona with its Tucson conferences, first established in 1994. In both cases, the research involved has been of necessity interdisciplinary; in both cases both materialists and transcendentalist have been involved. While science is still very cautious in straying beyond the bounds of strict causality and reductive materialism, artists are prepared to look everywhere and anywhere to try to solve this mystery. Science is caught in a trap of its own making: on the one hand, it recognises the counter–intuitive precepts of quantum physics, while refusing to recognise their metaphysical implications. In so far as altered states of consciousness are concerned, science is in denial.

Artists, untrammelled by orthodoxy (though no less concerned with authenticity), are prepared to examine any discipline, scientific or spiritual, any view of the world – however esoteric or arcane – any culture, immediate or distant in space or time, in order to find ideas or processes which might engender creativity. There is no meta–language or meta–system that places one discipline or world–view automatically above all others. Such transdisciplinarity can inform artistic research at all levels. This is why we look in all directions for inspiration and understanding: to the East as well as the West; the left hand path as well as the right; working with both reason and intuition, sense and nonsense, subtlety and sensibility.

In the early 20th century, science outlawed consciousness from its canon. For millennia before that, and certainly with the emergence of science in the 18th century, consciousness was openly discussed. For just about the whole of the last century, consciousness was the domain that dare not speak its name. It was David Chalmers as much as any one who, in 1994, opened what can be seen a credible bridge between the two opposed discourses of science and consciousness, with his identification of the hard problem. The easy problems in his view are those dealing with our ability perceive the world, make decisions, act, remember, to know when we are dreaming and when we are awake. On the other hand,

> *"The really hard problem of consciousness is the problem of experience. When we think and perceive, there is a whir of information processing, but there is also a subjective aspect. As Nagel has put it, there is something it is like to be a conscious organism. This subjective aspect is experience. When we see, for example, we experience visual sensations: the felt quality of redness, the experience of dark and light, the quality of depth in a visual field. Other experiences go along with perception in different modalities: the sound of a clarinet, the smell of mothballs. . . . What unites all of these states is that there is something it is like to be in them. All of them are states of experience".*

To state the hard problem is of course not to solve it, but Chalmers offers an approach that keeps open the possibility of consciousness being a field we inhabit (rather in the way that we inhabit space). His doctrine of psychophysical supervenience (first introduced into the philosophy of mind by Donald Davidson) amounts to "no mental differences without physical differences", without however ascribing to these physical differences an irreducible causal responsibility for differences in mental states. We know that these states of experience are limitless in their variety, some seemingly having their source on other worlds, but not just different universes of discourse – though they can alter powerfully our perceptions and sense of self – but new spaces of consciousness which we access through somatic or spiritual exercises, ritual ingestion of plants, sacred dancing, or the technologies of virtual reality and telematic communication.

We should perhaps be open to the idea that "individual consciousness" may be an oxymoron. While individual *self–awareness* is a prerequisite of living beings, consciousness is more likely to be the attribute of a field than of the individual organism. There is a certainly a tendency within modern science to see it this way: materiality as an epiphenomenon of consciousness rather than consciousness as an epiphenomenon of the physical body, as classical science has always insisted. A strong advocate for this point of view is Hans–Peter Durr, of the Max–Plank–Institut für Physic, Munich. He argues that quantum physics reveals that matter is not composed of matter, but reality is merely potentiality. His research suggests that the world has a holistic structure, based on fundamental relations and not material objects, admitting more open, indeterministic developments. From this it follows that in this more flexible causal framework, inanimate and animate nature is not to be considered as fundamentally different, but as different order structures of the same immaterial entity. In a stable configuration, effectively all the uncertainties are statistically averaged out, thereby exhibiting the unique and deterministic behaviour of ordinary inanimate matter. In the case of statistically unstable but dynamically stable configurations, the 'lively' features of the underlying quantum structure have a chance to surface to the macroscopic level and be connected with what we observe as the phenomenon of life.

From this point of view, evolution can be considered as a purposive pathway towards increasingly greater access to the field of consciousness, where survival is measured on a spiritual level: the fittest being those most able to adapt their individual self– awareness to the larger whole. This could temper the often aggressive interpretation of 19th century ideology of Darwinism with a more purposeful model of co–operation and collaboration, from the molecular to the macroscopic level of life, in the way that Lynn Margulis has argued, and as the research of Mae–Wan Ho supports:

> *Many remarkable individuals and local communities are indeed changing their own lives and the world around them for the better. They all do so by learning from nature and recognizing that it is the symbiotic, mutualistic relationships that sustain ecosystems and make all life prosper, including the human beings who are active, sensitive participants in the ecosystem as a whole.*

In the evolutionary process, which the scientific community may yet come to see as more teleological than wholly random, both telematic and pharmacological technologies can serve the transformation of the self and the connectivity of minds. In this respect I anticipate the eventual emergence of a syncretic alliance. In some cultures, "vegetal" technology, administered in a sacred and ritualised context, enables the navigation of altered states of consciousness, and a pathway to a spiritual domain. As much as it is at the root of many traditional cultures, immaterial connectedness confers a metaphysical dimension tokk both telematic art and quantum mechanics. Western nanotechnology, rather than simply being the end game of materialism, effectively provides a doorway to the quantum domain. Locative media and telematic communication put the mind out–of–body and globally distributed, altering the phenomenology of space and time. Technoetic research into states of immateriality and emergent materialisation may redefine our ideas of identity and presence. Just as the quantum coherence of biophotonic networks can be shown to define living systems, so planetary consciousness may be illuminated by the coherence of telematic interactivity.

Past ontologies have always taken the view that nature engineers events, rather in the way that the cosmos appeared to engineer time and space. Classically, we have assumed that time flows only one way, events are connected in causal relationships, and reality has been engineered *ab initio*. The clockwork model persists, with many cognitive scientists regarding the brain as leading in a strict causality to consciousness. This has led to the idea of reverse–engineering the process, with the fantasy of building an artificial brain that could "generate" consciousness.

Instead, a far more fruitful approach would be to design a machine that could *access* consciousness, seeing immaterial connectedness as prior to the materiality of the brain. Quantum science, while recognising the linear lawfulness of macroscopic events, takes the view that, at levels below its atomic and molecular materiality, nature is without the status of certainty; the complimentarity of particle and wave, matter and spirit, provides only for indeterminate outcomes. Non–locality and non–linearity are qualities at the very foundations of life, the specificity of all quantum events depending on our active participation as observers.

From this flow the following questions:

- Is our drive to created wider and deeper and faster networks an evolutionary impulse to engage more fully with universal mind?
- Does the telematic field of cyberception attempt to mirror or even augment our awareness of the field of consciousness?
- How aware are we of our own teleological promptings, the purposive impulse of our own DNA

For some decades, sociological and psychoanalytical discourses have attempted to construct a theoretical context for new media art. Often immured in a morose materialism they have failed to see the spiritual horizon, myopically setting their eyes on the "Other", limited to the gross level of reality. Now, as a science of consciousness arises, and as artists increasingly navigate its altered states, it is the metaphor of the Double that exercises our minds. In telematic space we are both here and out of body; mixed reality technology combines physical and virtual actions into a new kind of event space; the ingestion of entheogens allows us to move freely between worlds. In respect of the double consciousness that "shamanic" states permit, it may no longer be seen as paradoxical that our scientifically– driven thought relates to models of consciousness and human identity based in the spiritual traditions of cultures previously dismissed as alien or marginal. Art may increasingly take on a more psychoactive complexion, and it will be found useful to link archaic models of consciousness, such as we find in Amazonia for example, or amongst the Tsogho of Gabon, and ideas of quantum coherence, such as we find in biophysics, and biophotonic research. These archaic models implicitly locate the human mind within a field of consciousness, rather than seeing consciousness as an epiphenomenon of the brain, as western materialist orthodoxies would argue. Altered states of consciousness can be accessed by means of ritualized forms of breathing, dancing, chanting, or by the ingestion of psychointegrator plants. This understanding of consciousness as a field, and our ability to navigate it (and, as many aver, to be navigated in it by other spiritual entities) is seen most vividly in the syncretic doctrine of afro–Brazilian Umbanda, which brings together African Yoruba, and European spiritist beliefs with the native wisdom and traditions of the forest. Equally, from the Buddhist point of view, the mind is not a by-product of the brain, but a field that is a separate entity from the body, and which confers an inherent connectedness on the human condition.

Biophysics is a field-based science. Recently, field theories have been usefully sketched out in both their biophysical and metaphysical dimensions by Jean McTaggart in *The Field*, just as a field-based morphogenetic model of biological process and its spiritual implications informed Richard Sheldrake's *A New Science of Life* twenty years ago. Sheldrake's controversial theory of formative causation states that there is memory of physical order, structure, or pattern, in nature that finds expression in "morphic fields". The memory in these form-fields comes from previous forms of a similar kind. Morphic fields are an organizing principle of nature. He supports the contention that genes carry only a very small part of the biological information in a living system; most of it is in the memory carried within the organizing fields of an organism. Over time, the development of a larger memory of species experience, leads to the process of "morphic resonance", where at all levels in nature, the form of systems is influenced by the form of previous systems. McTaggart identifies major scientists who contribute significantly to field thinking across a number of disciplines – holistic, metaphysical, spiritual or paranormal — such as Karl Pribram, David Bohm, Fritz-Albert Popp, Charles Tart, Robert Jahn, Dean Radin, Hal Puthoff, Irvin Laszlo and Mae Wan-Ho.

Field thinking informs an understanding of healing practices of various kinds. Research into the connection between the biophoton parameters and the parameters of electromagnetic fields active on living system such as that undertaken at the laboratory of the International Institute of Biophysics at Neuss, Germany may provide some scientific validity to those ideas of self-regulation of the body to which various spiritual practices and somatic therapies subscribe. The network of "meridians" in acupuncture may be related to the body's biophoton field, as may the "prana" of Yoga. But the very inconclusive nature of scientific research in these areas, opens them, perhaps inevitably, to consumer abuse on the web, just as western medical jargon has long been exploited for the purposes of quackery, deception and commercial gain. However, just as the healing rituals in older cultures involved performative, interactive and imaging activity, it may be that art in contemporary society will come to acquire a more compelling value. In art, it is the field of interactivity that integrates the work, the artist and the viewer in what is both a material and an immaterial connectedness.

An organism's information network of photons that DNA molecules emit, may be seen as technologically paralleled by the instant flows of electrons and photons across the body of the planet through telematic networks.

Briefly to describe the basic proposition concerning biophotonic process, we can turn to Fritz–Albert Popp:

> *Biophoton emission is a general phenomenon of living systems. It concerns low luminescence from a few up to some hundred photons per second, per square centimetre surface area, at least within the spectral region from 200 to 800nm. The experimental results indicate that biophotons originate from a coherent (or/and squeezed) photon field within the living organism, its function being intra and intercellular regulation and communication.*

Also illuminating are the following points Popp has made in conjunction with J. J.Chang.

- Bioelectrical or bio electromagnetical phenomena have been known for a long time, but the coherent bio–electromagnetic fields, including biophoton fields are a new concept.
- They exist in living biological systems although we cannot see them.
- They are some kind of structure with specific patterns, but they are not real matter, only fields that regulate and bring the living system into a coherent state.
- In such a state within the coherent volume, there is no difference between particles and waves, therefore distance has no meaning.
- This state provides ideal conditions for the communication that is the basis for biological regulation.

Within quantum field theory, the coherence thought to define a living organism conforms to the understanding of quantum mechanics that holds that material reality forms an unbroken whole that has no parts. The reductivist world–view of classical physics must give way to the understanding in quantum mechanics of the primacy of the inseparable whole, and of the fundamental interconnectedness *within* the organism as well as *between* organisms, and that of the organism *with the environment.* This assertion finds support in the work of a number of leading physicists, amongst whom, for example, David Bohm and Karl Pribram independently arrived the understanding of the holographic nature of reality.

In the search for mind, especially in the context of the artist's use of technology to explore consciousness, the technologies of other cultures can provide an important example. As much as data is stored deep in the memory space of the computer, so knowledge is stored deep in the psychic space of the shamanic world. Western codes and protocols of computer access find their equivalent in the rituals and procedures of sacred ceremonies. In the old, traditional cultures, another technology predominates, providing its users with tools of consciousness and a spiritual technology whose use and history lies beyond historical record. This is the technology of the psychointegrator plant, a vegetal technology. Such plants as salvia divinorum or the shamanic liana, ayahuasca (banisteriopsis caapi), called the vine of the soul, and used in countless communities in Brazil and Colombia, are known as teachers, imparting wisdom as spiritual avatars. The researches of ethnobotanists such as Richard Evans Schultes, Eduardo Luna and Benny Shanon document the power of these plants in their sacred setting to enable us to transform consciousness, to enter into other states of being, to communicate over great distances, to connect with other entities, and to receive knowledge and instruction from the plant domain. In recent decades the use of vegetal technology to heighten spiritual experience has extended in towns and cities, in Brazil most extensively, but increasingly in other countries, largely through the practices of Santo Daime and União do Vegetal. The opening up of public awareness to the power of plants to heal the body and to transform the mind will doubtless infiltrate art theory, if not immediately the practice of art. Just as the artist's fascination with new technology has led to an electronic, interactive, telematic art, so it is possible to foresee a chemical or pharmacological ethos arising in art.

Indeed, it is my contention that the pharmacological processes of what can be called Vegetal Reality and the computational systems of Mixed Reality could combine to create a new ontology, just as our notions of outer space and inner space may coalesce into another order of cosmography. As can be seen in the work of Fritz–Albert Popp and other molecular biologists, DNA emits a weak form of coherent light that has been demonstrated to work like a communication system between cells and even between larger organisms. This suggests an information network of light not only within the body but also throughout and between all living things. It may not be too far to suppose that it constitutes the infrastructure of mind, accounting for the immanence of consciousness.

In the frame of ontological engineering, rather than in direct engagement with vegetal effectors, the West has pursued a more synthesising approach to the study of altered states of consciousness, using the science of chemistry to investigate the organisation of the brain and to provoke changes of emotion, perception and cognition. In *The Chemical Architecture of the Human Mind: Probing Receptor Space with Psychedelics,* Tom Ray reports on providing the first comprehensive view of how nineteen psychedelic compounds interact with the human receptome. Understanding the chemistry of consciousness is the final objective of this research.

Finally, taking these issues I have raised in their totality, it is my contention that the development of a truly technoetic art will emerge from the confluence of connectivity, syncretism, and field theory. Connectivity is at the root of cultural coherence, syncretism at the root of spiritual coherence, and field theory at the root of quantum coherence.

Bibliography

Ascott, R. 1980. Towards a Field Theory for Post–Modern Art. *Leonardo* (San Francisco),13, pp.51–52.

Bohm, D. 1980. Wholeness and the Implicate Order. London: Routledge and Kegan Paul.

Brown, D. 1986. *Umbanda: religion and politics in urban Brazil*. New York: Columbia University Press.

Chalmers, D. J,. Facing Up to the Problem of Consciousness, *Journal of Consciousness Studies* 2(3):200–19, 1995

Chang. J.J. and F.A.Popp.1998. Biological Organization: A Possible Mechanism based on the Coherence of Biophotons. In: *Biophotons* (J.J.Chang, J.Fisch and F.A.Popp, eds.), Dordrecht: Kluwer Academic. pp. 217–227.

Davidson, Donald, *Essays on Actions and Events* (Oxford: Clarendon Press, 1980)

Dennett, D C. 1992. *Consciousness Explained*. New York: Pantheon. Dennett could be described as the pope of epiphenomenalism.

Dürr, H–P. 2002. Inanimate and Animate Matter: Orderings of Immaterial Connectedness – The Physical Basis of Life. In: H.–P. Dürr et al (eds). *What is Life? Scientific Approaches and Philosophical Positions*. New Jersey: World Scientific. Pp 145–166.

http://www.lifescientists.de/ib_000e_.htm (accessed 16 November 2004)

Luna, Luis Eduardo & White, Stephen F. (eds.). 2000. *Ayahuasca Reader: Encounters with the Amazon's Sacred Vine*. Santa Fe, NM: Synergetic.

Mae–Wan Ho, 2000.*Towards a New Ethic of Science*.

Margulis, L. 1970. *Origin of Eukaryotic Cells*. Yale University Press.

McTaggart, L. 2003. *The Field: The Quest for the Secret Force of the Universe*. New York: Quill.

Nagel, T. (1974). What is it like to be a bat? *Philosophical Review*, 83(4):435–50.

Polari de Alverga, A. 1999. *Forest of Visions: Ayahuasca, Amazonian Spirituality, and the Santo Daime Tradition*. Rochester, Vt.: Park Street.

Pribram, Karl.1969. 'The Neurophysiology of Remembering', *Scientific American* 220: 73–86.

Ray, T. 2005The Chemical Architecture of the Human Mind. *Corante*. http://www.corante.com/brain-waves/archives/2005/02/05/the_chemical_architecture_of_the_human_mind_by_tom_ray.php (accessed 3 March, 2005)

Schultes , R.E. and Raffauf , R. 2004. *Vine of the Soul: Medicine Men, Their Plants and Rituals in the Colombian Amazonia*. Santa Fe, NM: Synergetic.

Shanon, B. 1999. "Ayahuasca visions: A comparative cognitive investigation," Yearbook for Ethnomedicine and the Study of Consciousness 8. Edited by C. Rätsch & J. Baker. Berlin: VWB Verlag.

Sheldrake, R.1983. *A New Science of Life*. London: Granada.

Winkelman, M. 1995. Psychointegrator Plants: Their Roles in Human Culture, Consciousness and Health. In: Winkelman, M. & W. Andritsky (eds). *Yearbook of Cross–Cultural Medicine and Psychotherapy*. Berlin: Verlag fur Wissenschaft und Bildung.pp.9–53.

www.consciousness.arizona.edu/archived.htm (accessed 15 May 2004)

www.lifescientists.de/ib0204e_1.htm (accessed 16 November 2004)

www.planetary–collegium.net/conferences/ (accessed 15 May 2005).

2. The Body
2.1 Are the Semi–Living Semi–good or Semi–evil?
Ionat Zurr and Oron Catts

The use of animals or parts of animals by humans has been practised since the dawn of our species. Selective breeding represented one of the major shifts in human society and culture. By employing the principals of hereditary traits humans transformed themselves into agrarian sedentary societies while transforming wild species into domesticated varieties that never existed before. Living organisms were appropriated by humans and modified/enhanced for functional and aesthetical purposes as the basis for agricultural practices (such as ornamental plants and fish, husbandry animals and pets). Much earlier, the use of fresh and preserved organisms (wood, leather, ivory, etc.) had been part of the human constructed world.

With the aid of our newly acquired knowledge of life processes – from ecologies to molecular biology – we can exercise an ever–growing degree of control over the manipulation of living biological systems to the extent that the technosphere ('human made') and the biosphere ('nature') are increasingly indistinguishable. The ability to cut and paste genes from different organisms, the prospect of designing artificial genes, and the possibility of coercing living functional tissue (outside of an organism) to grow and behave according to human determined plans, are just some examples of this merger. Artists are now exploring the new knowledge and tools offered by modern biology to manipulate and create living and semi–living works of art.

Are the kinds of manipulations offered by modern biology so different from the past ones? For one, Western urban society is different; therefore our relations with other living systems and their manipulations are different. The language we use to describe living and semi–living systems reflects our changing relations towards them.

Contemporary Settings
This paper is written in the light of the hideous rhetoric used in these pre–war times, and its relevancy to the creation of a new class of object/being, the Semi–Living, which are grown/constructed by TC&A. TC&A was initiated in Perth, Western Australia. In *The West Australian* newspaper we recently read about the commencing Australian campaign against terrorism offered by the Australian government as part of the new plan to combat global terrorism and to protect the Australian people (excluding the refugees who are enclosed in detention centres).

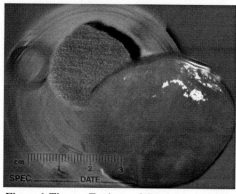

Figure 1 Tissue–Engineered Steak No. 1, 2000. A Study for 'Disembodied Cuisine'. Medium: Pre–natal sheep skeletal muscle, cultured and grown for four months, over PGA scaffold.

The extract below discusses the changes made to an Australian government advertisement campaign for a free toll number to report 'suspicious behaviour'. The campaign was launched on 30 December 2002:

To avoid alarm, the government has ditched images of balaclava–clad Special Air Service troops storming houses and police look–outs on the Sydney Harbor Bridge for the first phase of the campaign, in favor of softer pictures. Originally tested among focus groups, the pictures have been replaced by smiling Muslim girls, indigenous children and summer scenes of park cricket and barbecues, spliced with shots of army, Customs personnel and sniffer dogs at work.

Australians are friendly, decent, democratic people, and we're going to stay that way. Security agencies have been upgraded and are ready to detect, prevent and respond to terrorism. We can all play our part by being alert, but not alarmed. (Karen Middleton, *The West Australian*, 28 December 2002, pp. 1 and 10)

The same newspaper, on the same day announced that the Clonaid Company, established by the Raelians, had success in creating the first cloned baby: 'Sci–fi cult claims birth of first cloned baby' (p. 3).

Our new knowledge of biological sciences and its applied technologies enable human civilization to create new forms of life. Drawing on the example of the Raelians, there is enough variety and eccentricities in our own species to make it a reality. Though looking at the level of compassion to living systems of our own species, from different ethnicities, religions, races and class, we are worried in regard to these new lives. The difference between being alert and being alarmed is as illusive as the difference between a 'terrorist' or a 'freedom fighter'. Who is excluded/included from the definition of being an Australian? And is that reflecting them as friendly, decent, democratic people? We doubt that.

The form and the application of our newly acquired knowledge will be determined by the prevailing ideologies that develop and control the technology, which in turn will change according to new possibilities offered by the knowledge. Georges Canguilhem, referring to Darwin's observation[3] that variations in nature will not have any effect without the forces of natural selection asks: 'What could limit the ability of this law, operating over a long period of time and rigorously scrutinizing the structure, overall organization and habits of every creature, to promote good and reject evil?'[4] We might need to direct him to George W. Bush's speeches post–September 11 to find an answer to this question. When the manipulation of life takes place in an atmosphere of conflict and profit–driven competition, the long–term results might be disquieting. One role that art can play is to suggest scenarios of 'worlds under construction' and subvert technologies for the purpose of creating contestable futures, and exploring variations that might have an effect. This role of art makes the emergence of the Semi–Living and the multilevelled exploration of its use so relevant. Collections of cells cooperating/competing together for some sort of coherence that will enable survival are now being manipulated/exploited by us.

Evolution and cell theory: competition versus collaboration

'The introduction of cell theory in the biology, first of plants (around 1825) and later of animals (around 1840) inevitably turned attention toward the problem of integrating elementary individualities and partial life forms into the totalizing individuality of an organism in its general life form'.[5]

The very word 'cell' was coined by Robert Hooke, who was the first to observe the structure of cells in slices of cork. The structure reminded him of honeycomb. Canguilhem inquires,

Yet who can say whether or not the human mind, in consciously borrowing from the beehive this term for a part of an organism, did not unconsciously borrow as well the notion of the cooperative labor that produces the honeycomb? ... What is certain is that effective and social values of cooperation and association lurk more or less discreetly in the background of the developing cell theory.[6]

The dominant discourse/rhetoric surrounding evolutionary principles based on models of competition and struggle lead to the concept of the survival of the fittest. This ideology has originated in Darwin's own writing. Darwin's writing on the origin of species stemmed from the economic theories that were developed in the late eighteenth century. Adam Smith's *An Inquiry into the Nature and Causes of the Wealth of Nations*, which was published

Figure 2. *SymbioticA Central Lab–PC2. Part of 'BioFeel' Exhibition, Perth, Western Australia, August 2002.*

in 1776, argues for a natural basis for poverty and the need for a free market as a model for progress and innovation.

Evolution has no aim, it is not a linear process and it does not progress towards 'something'. According to Gould,

> *There is no progress in evolution. The fact of evolutionary change through time doesn't represent progress as we know it. Progress is not inevitable. Much of evolution is downward in terms of morphological complexity, rather than upward. We're not marching toward some greater thing.*[7]

The nature of the explanations of the mechanisms governing evolutionary principles reflects the dominant ideologies of our society rather than some scientific truth.

The microbiologist Lynn Margulis (1981) has offered an alternative emphasis in regard to the evolutionary process. She theorized that some of the greatest leaps in evolutionary development are caused as a result of cooperation and symbiotic relationships. She suggested that a eukaryotic cell is a result of the evolutionary symbiotic relations between two prokaryotic bacteria. Her theory was based on the existence of more than one set of DNA in the eukaryotic cell, one which originated in the germ cells and is found in the nucleus of the cell, and the others which are based in the protoplasm of the cell (which has some functions to do with the activity within the cell). In 1927 the American biologist Ivan Wallin (1883–1969) wrote, 'It is a rather startling proposal that bacteria, the organisms which are popularly associated with disease, may represent the fundamental causative factor in the origin of species'. Can it be that the basic building blocks of our own bodies, hence the eukaryotic cell, is a result of symbiotic relations between two entities (different bacteria)? Can it be that the origins for our own functioning body are collaborations between the entities we consider to be our enemies? Drawing on the Margulis theory we reveal the complexity of our definitions of what is 'Us' (our own body) and the 'Other' (the enemy). The rhetoric and the context of these elements in the fabric of human existence are determined by the ideologies they rose to serve. Good and Evil might seem as clear definitions to the President of the United States (and the Australian Prime Minister John Howard) but even in the context of what consists the evolution of our own physicality these definitions are blurry and contradictory.

> *By symbiosis different varieties of bacteria came together and made cells with nuclei. These cells with nuclei often cloned themselves into multiple copies that stayed in physical contact after reproduction ... But plant, animal and fungal life greatly expanded the complexity of the free–living protist cell by repeating it to make multi-*

cellular copies that ultimately evolved into separate tissues, such as reproductive and nerve tissue, with distinct functions.

By evolutionary processes, which largely depend on mutative incidents and adaptation to environmental factors, communities of eukaryotic cells have created collaborating communities that 'enabled' differentiations into varied specialized functions within one community. These communities enabled survival and immortality by sexual reproduction.

Figure 3. Harvest of pig mesenchymal cells (bone marrow stem cells). Part of the 'Pig Wings' Project 2000–01

TC&A is exploring the level of organized cell communities – the tissue as a palate for manipulation. Drawing on the collaborative nature of cells, we are growing these tissues separated from the body and coercing them to grow in predetermined shapes by the use of artificial scaffolds. We are also interested in employing their differentiated function for purposes other than what they were intended for. By that we are exposing and questioning long–held beliefs and assumptions in regard to the nature of things.

TC&A and the Semi–Living
Manipulations of a whole organism have been traditionally done as part of selective breeding practices. The interventions into living systems are now being perfected as part of molecular biology, which enable more precise and faster control over the manipulation as well as allowing cross–species mergers. Artists dealing with genetics consider the genetic code in a similar way to the digital code. As a result the manipulation of life becomes 'manipulation of a code', though with 'real' physical consequences that may appear in the phenotype of the manipulated organism. The manipulation of living tissues (from a complex animal) outside and independent from the animal they were derived from is possible by and dependent on other tools of modern biology – tissue and cell culture techniques as well as the emerging field of tissue engineering. From our own perspective, manipulations in the level of the tissue raises the most intriguing epistemological and ethical questions, as there is not yet an existing discourse that deals with the Semi–Living. Growing parts of an organism independent to it complicates notions of what life, self and identity are.

TC&A looks at the level above the cell and below the whole organism, hence at the collective behaviour of cells which forms tissues. We are using tissue engineering and stem–cell technologies to create Semi–Living sculptures. TC&A is introducing a new class of object/being in the continuum of life: the Semi–Living are sculpted from living and non–living materials, and are new entities located at the fuzzy border between the living/non–living, grown/constructed, born/manufactured, and object/subject. The Semi–Living relies on the vet/mechanic, the farmer/artist or the nurturer/constructor to care for them. They are a new class of object/being that is both similar and different from other human artefacts (human's extended phenotype) such as selectively bred domestic plants and animals. These entities consist of living biological systems that are artificially designed and need human and/or technological intervention for their survival and maintenance. As artists we are examining the Semi–Living as evocative objects;[8] prompting the re–evaluation of our perceptions of what life is and our treatment of other forms of life.

Our Semi–Living sculptures need to be kept in sterile conditions, immersed in nutrient media and kept at a temperature that suits their needs (mammalian tissue usually kept at 37 degrees, fish and amphibians can be left at room temperature). For their survival they need to be protected and fed on a daily basis. As a result of these needs, the exhibition of the Semi–Living sculptures in galleries (or other public spaces) involves new procedural acts and rituals as part of the artistic experience. In order to sustain the sculptures alive we need a tissue culture laboratory.

We have made a decision to incorporate the laboratory as part of the installation in order to present the environment in which such Semi–Living entities can thrive. A lab in the gallery also enable us to perform the duties needed to care for the sculptures in a manner that the audience can watch and comprehend the commitment and responsibilities we have to exercise towards the living systems we created. This involves the construction of an enclosure area, a sterile hood, an artificial environment for the Semi–Living sculptures (a bioreactor), laboratory consumables and the safety requirements of physical containment level–two laboratories. All are being examined and designed/constructed as integral to the conceptualization and theatrical intentions of the installation.

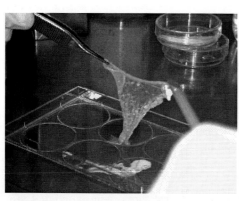

Figure 4. A layer of bone tissue differentiated from pig's mesenchymal cells (bone marrow stem cells) after four months of culture. Part of the 'Pig Wings' Project 2000–01.

Our installation involves performative elements that emphasize the responsibilities, as well as the intellectual and emotional impact which results from manipulating and creating living systems as part of an artistic process: *The Feeding Ritual* is performed every day. In this act we express notions of caring for Semi–Living sculptures, and invite the audience to view the process of feeding. The act that we refer to as feeding is being performed routinely in laboratories around the world and involves the replacement of used nutrient solution. At the end of every installation we are faced with the ultimate challenge of an artist – we have to kill our creations. Transferring living material through borders is difficult and not always possible, and as there is usually no one who is willing to 'adopt' the Semi–Living entities and care for them daily (under sterile conditions), we have to kill them. The killing is done by taking the Semi–Living sculptures out of their containment and letting the audience touch (and be touched by) the sculptures. The fungi and bacteria which exist in the air and on our hands are much more potent than the cells. As a result the cells get contaminated and die (some instantly and some over time). *The Killing Ritual* also enhances the idea of the temporality of living art and the responsibility that lies on us (humans as creators) to decide and act upon their fate. It is important for us, as part of the broader issues regarding life that we raise through our work, to expose to the public what happens to living systems artistically created at the end of the exhibition.

If the lab aesthetics should emphasize the nurturing aspects of life or their mortality, can it express both, as life without each one of these components cannot exist? At this stage of our project (and the scientific and technological abilities) human direct physical touch of the Semi–Living destined the Semi–Living to a process of death. The bacteria and fungi on the external layer of our skin contaminate the cells of the Semi–Living sculptures which lead to their demise. By the most obvious or 'natural' act of human nurturing, through caressing, we kill communities of cells which are stripped from their host body and immune system. When interacting with Semi–Living entities humans must learn to translate their limited understanding and perceptions towards a different set of instincts of a different living system.

The different treatments of the Semi–Living

Semi–Living sculptures are made out of parts of complex organisms that can only survive outside of a body with the aid of artificial support mechanisms.[9] Back in 1916, Edward Uhlenhuth, one of the pioneers in tissue culture declared: 'Through the discovery of tissue culture we have, so to speak, created a new type of body in which to grow cells'.[10] This 'new type of body' can be seen as a new life form. However, this new artificial life form is, at least at this stage; completely dependent on the good will of its operators for survival. The vocabulary associated with the Semi–Living entities is of crucial importance for its well–being. The irrational rhetoric leading to the next big war forces us to realize that we, humans, can be seen as 'tissue of horrible and disgusting absurdity'.[11] Though as 'tissue artists'

we found our tissue sculptures rather humble, collaborative and dynamic living communities that are in need of care.

The Semi–Living sculptures have no immune system and at this stage need to be protected from the external environment. One of our aims is to be able to grow an external protective membrane layer, rather like our own natural skin that will enable the Semi–Living to be exposed to the external environment and will enable a direct and tactile interaction with the audience.[12]

Up to date we have grown/sculpted the Semi–Living in different shapes, forms and tissue types and encouraged symbolic interaction with them. For example, in the Tissue Culture & Art(ificial) Wombs installation (firstly presented in 'Ars Electronica' festival 2000) we grew Semi–Living Worry Dolls.

Figure 5. Semi–Living Worry Doll H. Part of the Tissue Culture & Art(ifical) Wombs installation, 'Ars Electronica' 2000. Medium:McCoy cell line, and biodegradable/bioabsorbable polymers (PGA, P4HB) and surgical sutures.

The Worry Dolls are based on a Guatemalan tradition in which kids tell their worries to the dolls, then put them under their pillow with the belief that the dolls will take the worries away. We have grown Semi–Living Worry Dolls inside a bioreactor. Alongside the bioreactor we put a computer or 'a worry machine' to which people were encouraged to write their worries to the Semi–Living Worry Dolls. We have promised the audience that we will whisper their worries to the dolls with the hope that they will take their worries away. By logging on to our website you can view people's worries from all around the world (and add your own). This document is also a reflection of current anxieties; some are in regard to our own work, the ethics of biological art and biotechnology in general as well as in regard to life in an era of increasing global conflicts.

In a later project, 'MEART' aka 'Fish & Chips',[13] we began to 'demand' our Semi–Living entities for some information, tapping into the hazy area of sentience. Growing neural tissue (the 'thinking' units) and retrieving data from it, we have created what we refer to as 'A Semi–Living Artist': we have picked up electrical activity from neural tissue and transferred it to a computer program that drove a robotic arm and manipulated a musical score.[14] We used a sound–sensitive switch to control the audio output as a source of electrical stimulation feedback to the neurons. Working with communities of interacting neuron cells was an emotional experience. Our understanding of neural tissue as a 'thinking' unit, and as the place where consciousness resides made its manipulation more difficult. Questions in regard to the 'understandings' and sentience of the tissue that we hardly understand but yet manipulate made it ethically challenging. Epistemologically, the idea of future 'intellectual' communication with a neural tissue, which is grown independently from a body, raises many inspirations for a better understanding of the different levels of life.

Living and Semi–Living systems as food – 'Disembodied Cuisine'

Our latest project titled 'Disembodied Cuisine' explores another way of treating/interacting with living systems – by consuming them as food. Humans, like the rest of the animal kingdom, have always practised this ultimate way of exploiting other living systems. Though, as human society becomes urban and direct relations to what is considered 'wild nature' weaken, this behaviour is being further questioned. Furthermore, as our understanding of life increases, we employ different attitudes and hypocrisies to be able to continue this need (one must not forget that vegetables are also living systems). We recently heard a story from Jason Davidson, an aboriginal artist, who documented his hunting trips as a way to explain to his community about the functions of internal organs. He presented one of these videos to a white urban audience. In this video he showed a wild water buffalo that had been hunted and cut open for the dual purpose of food and education. One of the viewers could not hide her disapproval any longer; she accused him of being cruel, and suggested to him that next time he should go to the supermarket and get his meat there. This story epitomizes the hypocrisy of the Western urban society in relation to meat consumption. These neatly packed parcels of meat on the supermarket shelf bear very little reference to its source. One can argue that the act of buying meat in the supermarket, which is produced by growing animals in crammed industrial farms, is much more cruel than hunting (for food) an animal that had a good life in the wild. In the project 'Disembodied Cuisine' we have decided to further explore these relations with living systems by looking at the possibility of eating Semi–Living systems.

Semi–Living systems are a new construction/creation, made possible by the use of the tools of modern biology and which 'disobey' evolutionary processes. Furthermore, using Semi–Living systems as food enables us to eat parts of the animal while the animal is still alive, complete and healthy (after a short process of healing). There is also the possibility of creating non–pathological immortal cell lines for this purpose, and by that producing victimless meat. These new possibilities create an opportunity to discuss the relations between humans and their living food from a different and refreshing angle.

We have developed the project during 2000–01, while we were research fellows at the Tissue Engineering and Organ Fabrications, Massachusetts General Hospital, Harvard Medical School. We first published our intentions in an interview to the *Washington Post* in December 2000. We have used pre-natal sheep cells, harvested from a sheep uterus, to grow steaks the size of a coin. Since then, NASA have been experimenting in growing tissue-engineered fish skeletal muscle for food consumption in space travel.[15]

According to Claude Levi-Strauss, throughout history humans have instituted a strict division between what can and cannot be eaten. However, these divisions are not always clear, and we must practise some kind of hypocrisy in order to be able to love and respect living things as well as to eat them. Dogs are an example of such confusion; in some cultures they are 'man's best friend' (pets), and also ornaments that are selectively bred for aesthetic qualities. Dogs in other cultures are being eaten. Peter Singer refers to such division as 'Speciesism in Practice – Animals as Food'.

'Disembodied Cuisine' deals with one of the most common zones of interaction between humans and other living systems and will probe the apparent uneasiness people feel when someone 'messes' with their food while exposing the inconsistency of the ethical framework our society set up for dealing with non-human living systems. It can be argued that this inconsistency in turn plays a part in the double standard so apparent in the war rhetoric of George W. Bush and John Howard. In 'Disembodied Cuisine' the interaction/relationship with the Semi-Living is one of consumption and exploitation. However, it is important to note that it is about 'victimless' meat consumption. As the cells from the biopsy proliferate the 'steak' in vitro continues to grow and expand, while the source, the animal from which the cells were taken, is healing. Also, by that we may eliminate some of the problems associated with eating a whole sentient animal and with the way animals, destined for the meat/poultry industry, are being treated. However, by making our food a new class of object/being – a Semi-Living entity– we risk making the Semi-Living into a new class for exploitation.

In 'Disembodied Cuisine', which will be installed and performed as part of the international biological art exhibition 'L'arte Biotech', in Le Lieu Unique, Nantes, France from 13March 2003, we will attempt to grow frog skeletal muscle over biopolymer for potential food consumption. A biopsy will be taken from the animal that will continue to live and be displayed in the gallery alongside the growing 'steak'. This installation will culminate in a 'feast'.

Poppy Van Oorde Grainger, one of the students in the Vivoart class which is run by SymbioticA, the Art and Science Collaborative Research Laboratory in the School of Anatomy and Human Biology at the University of Western Australia[16] has offered a new twist to human confusion between a 'living or semi–living system' and 'a piece of meat'. Poppy, a vegan who believes in minimizing harm to animals, has confessed to us that recently she has had the urge to eat meat (can it be an 'evolutionary' desire for protein–rich food or the thrill of the hunt, or is it just an aesthetic desire for a different taste and texture of the food we decide to consume?). This desire for meat and the belief in not eating other species have found an outlet based on the idea of Semi–Living food. Poppy suggested taking a biopsy of her own body, rather than injuring an animal and inflicting physical and psychological pain (even if temporary) on another animal. By that we will be able to grow for Poppy steaks made of her own flesh. The questions we were pondering were not if it is against nature (we are a long way away from nature for a long time). Also humans have practised cannibalism before, nor if it is moral (it is done by a full consent from an aware adult), but rather questions of biosafety and furthermore, the rhetoric that will be used by our society to deal with such a concept.

The uneasiness

Why do we feel uneasy and even threatened in regard to the manipulation of life outside of the bounds of the evolutionary rules? As mentioned before, biological evolution cannot be described as a linear progress towards something better, stronger or wiser. However, human intervention in evolutionary biological processes is usually done 'in the name of progress'. Humans are accumulating better control; though not necessarily a better understanding of the long–term results of such interventions. In many ways we are not smarter than a cell or bacteria, and we can learn about our behaviour from the building blocks of our own bodies. The use of collaborative colonies of cells outside of a body is epistemologically and ethically a very relevant artistic expression which forces us to look at human civilization and its shifting rhetoric from an alternative position. Learning about communicative cells in a new 'unnatural' environment is like shining a mirror at our own behaviours.

Figure 6. Semi–Living Worry Dolls fixed in formaldehyde. The botched experiments towards part of the Tissue Culture & Art(ifical) Wombs installation, 'Ars Electronica' 2000. Medium: McCoy cell line, and biodegradable/bioabsorbable polymers (PGA, P4HB) and surgical sutures.

We feel extremely uneasy in relation to the rhetoric used by our own government. Drawing on the quotation from the Australian campaign to combat global terrorism, in the beginning of

Figure 7. Title: 'Killed' Semi–Living Worry Dolls as part of the Killing Ritual. 'BioFeel' Exhibition, Perth, Western Australia, August 2002. McCoy cell line, and biodegradable/bioabsorbable polymers (PGA, P4HB) surgical sutures and glow–in–the–dark plastic coffin.

this paper, we are asking: Do smiling Muslim girls represent a twisted mirror–reflection of balaclava–clad Special Air Service troops storming houses? Do the Semi–Living entities represent a dignified living system or a piece of meat for us to exploit and consume?

Growing Semi–Living entities might seem at first sight as technological determination or as going against nature. However, our motives are based in exposing social hypocrisies in regard to what is natural and also the shifting definitions of the 'other'. If we are not able to be compassionate for differences in our own species, will the existence of the Semi–Living or a collaborative symbiotic collection of cells enable us, even a little bit, to present a mirror of our absurdities? Can the naturalish qualities of the Semi–Living act as a surrogate to the 'real' nature that seems to vanish from the lives of urban humans? Is that an answer for Edward Wilson's Biophilia?

The Semi–Living entities complicate notions of life and of self as opposed to the 'other'. They are forcing us to realize that our cultural norms and values are ill–equipped to deal with the new knowledge in biology and the new creations made possible by biotechnology. Any technology is a double–edged sword and the uses of technology rely on complex interdependent relationships with the dominant ideologies at the time. Therefore, looking at the prevailing ideology of our times we are examining our Semi–Living sculptures as evocative entities, which offer tangible alternatives and contestable futures.

There are many issues that have to be resolved by humanity as a whole before we can proceed with large–scale exploitation of modified/designed living biological systems. This is of grave concern as decisions which are being made now will determine the directions in which exploitation of living systems take. It is of particular concern as we are entering an era of conflict and intolerance to the other, coupled with an extreme form of capitalism and profit taking. Artistic inquiry as the creation of Semi–Living sculptures generates engagement with these issues.

References

Ferdinand, P. *Washington Post*, 29 December 2000, p. A03.

Wilson, E. *Biophilia* (Cambridge, MA: Harvard University. Press 1984).

Singer, P. *Practical Ethics* (New York: Cambridge University Press 1993), p.62.

Delaporte, F (ed.), *A Vital Rationalist: Selected Writings from Georges Canguilhem* (New York: Zone Books, 1994), pp. 84–85.

Margulis, L. *Symbiosis of Cell Evolution* (San Francisco: W.H Freeman and Company, 1981).

Wallin, I. *Symbionticism and the Origin of Species* (Baltimore: Williams & Wilkins, 1927), p. 8.

Margulis, L. and D. Sagan, *What is Life?* (Berkeley and Los Angeles: University of California Press, 2000), p. 135.

Notes

1. For more see the Tissue Culture & Art Project website: http://www.tca.uwa.edu.au
2. By the time this paper will be published, the situation of 'pre–war' might be changed to a 'war' or 'after–war' situation. We are aware that we can use the term 'war situation' on a continuous basis as in our current global world there are continuous localities of war situations. However, this paper refers in particular to the 'war against terrorism' and the war against Iraq.
3. See *The Origin of Species* by Charles Darwin, chapter 14.
4. Georges Canguilhem cited in Francois Delaporte (ed.), *A Vital Rationalist: Selected Writings from Georges Canguilhem* (New York: Zone Books, 1994), p. 211.
5. Georges Canguilhem cited in Francois Delaporte (ed.), *A Vital Rationalist: Selected Writings from Georges Canguilhem* (New York: Zone Books, 1994), pp. 84–85.
6. Georges Canguilhem cited in Francois Delaporte (ed.), *A Vital Rationalist: Selected Writings from Georges Canguilhem* (New York: Zone Books, 1994), p. 162.
7. Stephen G. Gould cited in 'The Pattern of Life's History', *The Third Culture* (New York: Simon & Schuster, 1995), p. 52.
8. The term 'Evocative Object' coined by Professor Sherry Turkle, originally in regards to computers and other E–toys. For more see: Turkle, *The Second Self: Computers and the Human Spirit* (London: Granada, 1984).
9. Parts of an organism can be sustained alive for a long period of time (and in the case of cell lines – forever) while the organism from which the cells were taken from can cease to live. Furthermore, the biomass of the Semi–Living can be larger than the biomass of the original host.
10. Criticism on Mary Shelley Frankenstein, cited in Susan E. Lederer, *Frankenstein, penetrating the secrets of nature: An Exhibition by the National Library of Medicine* (New Brunswick, NJ and London: Rutgers University Press, 2002), p. 8.
11. By that the Semi–Living entities will become part of the environment, though without the ability to sexually reproduce.
12. For more see: http://www.fishandchips.uwa.edu.au
13. This project was developed by the SymbioticA Research Group: Guy Ben–Ary, Phil Gambleb, Dr. Stuart Bunt, Ian Sweetman, Gili Weinberg, Oron Catts, Ionat Zurr and Matt Richards.
14. For more see Wendy Wolfson, *New Scientist* 21–22 December 2002, pp. 60–63.
15. Hannah Landecker, *'Building "A new type body in which to grow a cell"*: Tissue Culture at the Rockefeller Institute, 1910–1914' in Creating a Tradition of Biomedical Research, edited by Darwin Stapleton, (New York: Rockefeller University Press) In press.
16. For more see http://www.symbiotica.uwa.edu.au

2.2 Absent Body Project

Yacov Sharir

The seductive power and agency of interactive art and 'virtual reality' has stimulated my imagination as a researcher and as an artist/performer. My disembodied self is re–embodied in cyber–bodies occupying increasingly immersive cyber–worlds, experiments of interactive art and wearable computer/devices. The self–descriptive, self–reflexive and recursive processes of consciousness reveal themselves as a dance of real and virtual, flesh and re–configuration, sensory presence and re–presentation, cognition and re–cognition. In the ordinary flow of conscious experience, these pairs are not encountered as binary oppositions in conflict but in a continual dance of transformation, one into the other. I converge with my own creations –the technological tools –and give birth to new tools and new gestures of consciousness. The research interests and my artistic practices have converged –literally and figuratively –in this zone of postures, gestures, movement and communication between real and virtual worlds and the effects on consciousness of such spatial practices.

A gestural human sign language and human postures are being exploited utilizing new possibilities through the use of new technologies as medium for inscription. Animated cyber–human characters/performers move, deform, rescale, rearrange themselves, augmenting the dimensions of expressiveness and meaning related to performance issues. The cognitive approach produced by thinking in and/or out of these bodies has similarly triggered changes in consciousness, thus affecting content and virtual story telling.

Together they explore the interaction of gestures as movement material –as in dance, human day–to–day gestures, postures and the capacity of carefully composed dance material to embody and generate meaning. Cyber dancers use gestures in order to cybernetically inscribe them. The composed gestures and movement material become a source of intention that relates to itself; its communicating environment becomes a visualization of the self–reflexivity inherent in the workings of both the dance and consciousness. The dance is between worlds of humans, cyber–humans, and the source language as it transitions and transforms into the domain of visible thought.

Methodologies include the human body and several interactive systems employing or wearing a wearable computer/device that is subsumed into the personal space of the user, controlled by the user and has both operational and interactional constancy, i.e. it is always on (if needed) and always accessible while in performance. Most notably, it is a device that is always with the performer and the performer can always enter commands into it while walking in and around the performance space. In my latest works it is used to enable a seamless multidimensional expression of intention and navigation through direct gestural interaction within a remembering knowledge space by way of accumulation. The working hypothesis for this aim is that a vocabulary of direct gestural expressions of creative intention can be recognized so that a choreographed set of gestures/movement can map the experiential body state of each gesture to corresponding system actions. The result will be an expression set, instead of a command set, through which the interactor (the wearer) communicates, rather than controls, the system.

Background

I am currently engaged in the creation of a series of works that investigate how I could conceive a choreographic system that will arrange and rearrange itself by itself and/or for itself —one that will not adhere to the guidance of logic but rather to a pre-conceived code producing non-linear hypertextual content that will provide structure and allow the work to progress gracefully with no beginning or end. Decisions that will be made by the wearer/user will adhere to specifically designed human movement gestures, postures, emotions and feelings through a gestural recombinant system. This system will not respond to traditional computerized commands.

My aim for these sets of works is that they take on a life of their own, recognizing their community of cyber performers/dancers, continually examining their ageless bodies and superb condition, bodies that will stop at a specified magnification of desired size, speed and astonishing liquidity and grace. The bodies will be placed against or adjacent to each other, they will gracefully and naturally 'by way of being' defy gravity and introduce a whole new vocabulary of raw movement material that is yet unexplored. They will take on a life , 'a way of moving', that is organic in nature and can co-exist (by choice) in peace and harmony with the physical human counterparts generously attending to issues related to how mutual performance space can/should be shared with each other.

'The Absent Body Project'[1] is a computer generated cyber human dance that exclusively employs computer-generated characters/performers stored on a mother ship that is placed strategically in the performance space. Very clean and clear computerized animated images/cyborgs were created with illuminated passages and transitions, a kind of work that does not need a linear plot but, rather, inhabits a browser of sorts or a search engine attempting to search for the performers' past experiences and or knowledge. This work did not need to be choreographed since it has neither a beginning nor an end. These characters are activated by/from a wearable suit/computer worn on a physical human performer

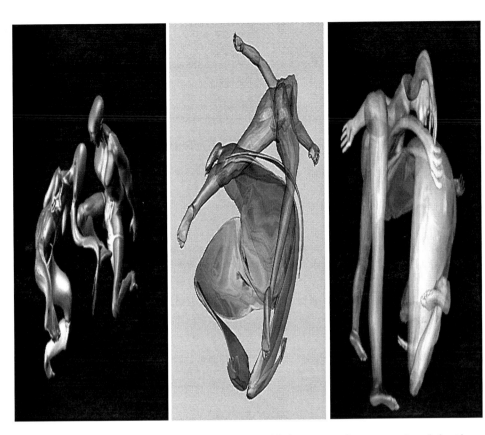

(myself). They are activated in real time by utilizing a set of commands originating, in this case, as hand movement material, 'a way of remembering'. These characters are commanded to operate in a structured, improvised and random way by the wearer; they can be activated to move forward or backward, to go fast or slow, to start or stop, to fast–forward and or rewind as needed in the moment. Additionally, the wearer/user can select to move instantly to the next set of characters/material that is waiting to attend performers. Thus, the dance is one of a kind sequence as it is experienced in the moment.

My Interest

My overall research/work/interest lies in the following four areas: the use of multiple interactive systems, the creation of wearable computers/devices, the design of animated characters/cyber–human performers, 3D worlds and virtual reality/environments. These research categories are created for the sole purpose of monitoring, controlling and augmenting all aspects of production/performance in real time.

My technological exploration is mediated through the use of choreographic works of art, human gestures and postures that form creative intentions, interactive visual language and cognitive linguistics that fuel creative innovation

Computers typically respond to our direct commands for well-formed actions but not to our spontaneously evolving inner desires and emergent ideas.

My way of working suggests that gestural recombinant knowledge and space visualization is a mean for augmenting creative innovation in a performative situation. My creative process is supported by a mode of textual human movement that is recognized by and through the use of multiple sensory devices and motion detectors. These devices then translate the material by numerical orders and/or by algorithm customization program(s) and mapping processes.

As the founder of the American Deaf Dance Company and its first artistic director, I am fluent in American Sign Language and possess great command over the way humans communicate with each other by utilizing an alternative language. Hands, arms, fingers, body postures, facial expressions and signs are mobilized for this important purpose. This language has served most of my technologically mediated choreographic works, providing me with the unique seductive power, agency and the movement linguistic that continue to fuel my imagination.

The complex layering of language, image and communication between worlds, both human and cyber–human, visualizes the shifting strata of memory, dream and conscious visualization and meaning–production. The technologically supported feedback that loops between the real and the virtual has produced profound shifts in consciousness, experiences of being both embodied and disembodied at the same time, 'a duality of existence'.

Wearable interactive devices

Wearable interactive devices and systems are in most cases subsumed into the personal space of the user, controlled by the user and/or placed in a desired location where the performer can activate them as needed. They possess operational command systems when placed on the physical body of the user/performer and interactional constancy when used as an interactive element, i.e. while used in performance they are always on and always accessible. Most notably, these devices have become an integral part and an extension of sorts of the user body. The user can always enter commands into it while operating in and around the communication area.

By using/exploring the use of interactive technologies in performative conditions I am seeking to discover how electronic and sensory devices affect the way we communicate, how we can alter the mind, zone of awareness, feelings and the walking consciousness. Interactive systems are much more than just wristwatches, jewellery devices, regular eyeglasses, different types of floor pads, and/or motion tracking devices. They possess the full functionality of a computer and a wireless communication system and are fully interactive. They are also inextricably intertwined with the wearer either when placed on the physical body, on the performance floor or when it functions as a motion tracking device/system.

Continuous research facilitates the augmentation of these systems' operational sensory devices and attributes so they can fully function as a medium for inscription. They accept/detect human moves, morphing, scaling, making colour changes and adding new dimensions of expressivities and meaning to performance. As for members of the public/the users, these systems provide them with the ability and opportunity to interact and manipulate the display of information/images both virtually and physically. Images can be displayed onto large screens subsumed into the performance/installation space. The ultimate intention is to surrender total authorship to the user and it is expected to be highly engaging, playful and somewhat provocative.

Considering my past experiences in the creation of virtual reality works such as 'Dancing with the Virtual Dervish'[2] and my current preoccupation with the creation and use of wearable computers and interactive technologies the 'Automated Body Project',[3] I am planning to explore/research in the following areas/ways: How can we spontaneously detect evolving inner desires? How can we consciously form creative and fluid intentions and find the ways/means for interactively expressing these intentions? What kind of an interface can be created that can embody the creative, the informal, the fuzzy and the physical?

The methodology suggested for such alternative research subjects/new work/process is as follows: An invisible/virtual door space 'a means/metaphor for entering the creative process' is suggested/proposed; an interface of sorts will be created. Participant enters, sees and or experiences choreographed gestures and mappings that can recognize different degrees of intensity for single or multiple gestures. The used space is visualized through a vision–based gesture organizer and video motion captures (front and side views) using the record mode for the gesture recognition program. Eyes see simulations of creative intentions and a model of parameter adjustment and content is facilitated.

Together they explore the interaction of gesture as language –as in dance, sign languages and mudras –and the capacity of an inscribed language itself to embody and generate gesture. Human and cyber–dancers create and use the physical gestures and cybernetically inscribed movement with the intention of communicating amongst themselves. The visual language itself models sentience, having the technologically mediated property to absorb or sense meaning and then communicate, both within and to itself, and outward to its environment.

The gesturing or moving and its communicating environment become a visualization of the self–reflexivity inherent in the workings of both language and consciousness. I demonstrate and report on this dance between worlds of humans, cyber–humans and language in the transformational domain of visible thought. The technologically mediated feedback loops between the real and virtual have produced profound shifts in consciousness, experiences of being both embodied and disembodied at the same time, both human and cyber–human, visualizing the shifting strata of memory, dream and conscious visualization and meaning–production.

Conclusion

As I am heavily immersed in an ongoing process of technological exploration, serious questions emerge suggesting topologies of temporality. Questions arise from the physical body, resisting automatization of sorts and embodiment as machines and at other times flirting with the subversive qualities of suggested magnified performance and the pleasurable pain it elicits. No matter how deeply involved I am in this process the major issues and questions remain the dramaturgy of performance content, internal and external time–consciousness, when–when is/was/will–be no longer, or at least differently, I am thinking–I think–I am–thinking I...

Notes

1. 'The Absent Body Project' was first performed by the Sharir+Bustamante Dance Works at the University of Texas at Austin, McCullough Theatre, 2003.
2. 'Dancing with the Virtual Dervish' was commissioned and first performed at the Banff Centre for the Arts, Canada, May 1994. It was also performed as an installation at the National Museum of Contemporary Art in Athens, Greece, 'Synopsis 2 –Theologies' 2002.
3. 'The Automated Body' was first performed by the Sharir+Bustamante Dance Works at the University of Texas at Austin, Oscar Brockett Theatre, 2000.

2.3 Our Body as Primary Knowledge Base

Kjell Yngve Petersen

When our abilities are augmented and multiplied, and our access to the world rearranges fundamental relationships, how then are we as body and entity transformed?

How can we understand what we experience and what we conceive when basic relationships have been extended and made fluid?

Are we becoming aliens, desperately trying to find new models or frames of perception of the world so we are able to comprehend a world which itself has become alienated?

No matter how advanced, fast, detailed or life-like these technological reflections might become, they still need our body in order to exist. There is no virtual world without a real person experiencing it; there is no extension of the senses or the actions without real senses and actions to extend. There is, in fact, nothing without the presence of the body. It is through the body that we have a place in existence. And it is through dynamic interrelationships, through actions and the senses, that the notion of reality comes about, and thereby our notion of our own existence.

I will discuss how the technological augmenting of our access to the world can be understood from the perspective of the body, if one investigates these new conditions using the methodologies and theories from advanced formal body language.

It is a method that uses the performer as a 'super-' or 'extra-' human tool to investigate through action and analysis. It is a way to test how realities occur under fluid and dynamic conditions within a controlled environment of time and space. It is a way of exploring possible constructs of 'human realness' detached from the chaotic circumstances of real life.

I propose to use this knowledge, free it from the purpose of training and performing, and develop and use it as a tool by itself –developed to a level where it becomes self-reflective and self-contained, evolved to a complexity where it becomes an entity by itself –and therefore can be a tool for something else.

It is a formal 'technical' use of the body as an extra-human behavioural entity. This involves the whole complex spectre of intellect, emotions, desires, memory, actions, etc. as a formalized 'machine'. To enable meaningful constructions of telematic and augmenting technology, one can use the formal methodologies of advanced formal body language as a tool in the creation process and thereby have 'the actual experience' present in the process as a monitor and constructive tool.

The creation of a dynamic human tool

This is knowledge of how to use your own body as an instrument. As such, it can be shaped in almost any way needed. It is specifically suited to deal with dynamic relations, the flow of interaction, relationships and intentions.

I will give a few examples of the process of preparing the body to be a senso–motoric entity, which then can be used to monitor and validate experience and at the same time be a creative tool in the design of telematic, performative, interactive installations and performances.

It is basically different physical and mental concepts that free the body from purposeful demands and make it an entity for pure experience and action. The process of preparing the body to be a senso–motoric entity is a well–known concept to many of the different performative genres of dance, theatre and performance. I will show how the use of the concepts and methodologies normally fixed to specific aesthetic expressions can be developed to be a tool by itself and therefore be useful in other art–genres and for other artistic purposes.

Here are a few examples of how the body can be thought of and transformed into a controllable tool for artistic expression:

1) It could be the control of the position and the form of the energy/power in the body that defines the mode of presence. In Japanese art forms this is placed very low as a strong tension giving force to a very controlled and powerful presence. In Indian art forms it is defined as an elevation or dance along the spinal column, giving a rhythmic and very active presence. In European abstract mime it is moved around to support a variety of figurative constructs. The control of the position and the form of the energy/power in the body can be used, in a formal way, to define the basic presence, and to some extent, the nature of the temperament, the emotional attitude and the energy of the intention. This establishes a formal concept of an intentional being.

2) It is a way of reorganizing the muscular tensions so that any movement is a controlled combination of contraction and release. It is a reorganisation of the muscular resting–points by merging the action and its counter–action into one state of possible action. It makes one able to perform or redirect any action from any position with equal effort and thereby also measure any resistance and relative distance.

3) It could be the build up of a formal distance to suit the purpose of the actions. This will establish and extend the mental space between a wilfully controlled action and giving in to being out of control, between determination and indecision –which establishes the quality of hesitation. It is the development of a space of time between action and in–action, in which degrees of meta–action can emerge. The intentional act emerges when the ordered past meets the present in the action of expectancy. The intentional act in the time–space of hesitation generates a possibility for articulation. This enables one to work with and be sensitive to subtle variations and resonance in expressions and impressions.

An extensive and detailed introduction to the various performance traditions and their methods of preparing and using the performer as a tool can be found in Eugenio Barba and Nicola Savarese, *The Secret Art of the Performer*. The knowledge of the physical practices of the performer is implemented and expressed differently in different traditions and cultures around the world but, as shown in the work of Theatre Anthropology, there is a basic concurrent knowledge independent of cultural differences.

This is a kind of scientific tool–making out of the human body/entity, transforming the 'human–being' into the 'human–tool'. We all do this to a certain extent as ordinary human beings –having different complex abstract distances to ourselves or qualifying ourselves to gain better tool–like abilities –this is just a very rigorous use of this inherited ability –bringing these abilities to a level where they become entities in themselves. This is not a de–humanization or alienation of the human –it is more a heightened, extra–human state. It is not an emptying of personality but a construction of formal distance to oneself as an entity.

Telematic performative environments

Performative telematic environments are technological constructs where 'human realness' is shaped and transformed by the technological rearrangement of our relationships.

Telematic augmentations are quite difficult to conceive. As soon as you have just a few telematic operations going on, they exceed the simple relations of cause and effect and of mere transmissions. They easily involve more adjustable parameters than one can comprehend. They become montages of non–parallel and not directly related occurrences of sensing and acting, representing a multiple of times, places, expressions and impressions. They often blur, alter, mix or augment our normal possibilities and therefore reveal to us how fragile, inexact and complex the normal interpretation of our senses and actions are. They ask for a reconfiguration of the construction of the 'human real'.

The telematic infused artworks are without the normal limitations and constrains that we humans usually experience in the physical world we know. There is no built–in hierarchy, behavioural tendencies or phenomenal distinctions –it is precisely a multi–ability/purpose technology.

There are a vast number of variable parameters and possible connections, relations and transformations, making it more of a problem to reduce and numb the telematic system to form comprehensible constructs than to develop technological refinements.

Telematic, performative artwork can successfully be progressively formed and tested in test set–ups where everything is scaled 1:1 and in real time. By designing the creation process with constant real–size, real–time testing and shaping, one can involve the totality of variables in their right relative proportions and through this promote coherent design solutions. It is the design methodologies used in most performative art forms –and I think the tasks at hand in telematic, performative artworks are much the same. They function in a relationship to the human being as a totality and are mainly concerned with dynamic and cybernetic phenomena.

These technological constructs are made for a human–dynamic experience. They are made to promote telematic presence, action and relationships and thereby evoke situations for passion, emotion, seduction, tension and provocation. But the constructs themselves are only dumb tools, which can supply human beings with advanced extensions and an environment of communicating artefacts. This generates a situation very similar to the training situation of the performer in the laboratory workspace, where the performer's body is reshaped to be able to generate that specific artistic multi–functionality which is requested in each specific performance. It is a tuning of the body to a specific instrument, which then is played on by the performers themselves while acting –in a kind of extra–human state.

In the construction of telematic, performative artwork one can use these 'extra–humans' as 'any–body' or 'some–body' –as specially designed 'people' representing the average visitor in a controlled way –or sharpen them to focus on only one parameter or relationship as a dedicated tool. To deal with the complexities of telematic, performative artwork, which is supposed to work with 'anybody', the most obvious approach is to build it using 'somebody' –a human being in a state of heightened awareness and with detailed skills to monitor and act in dynamic situations. This is an entity dedicated to conceive the circumstances of the telematic structure in a bodily way or as a field of presence.

One can look at telematic, performative artwork as conditions for possibilities and new kinds of spaces. They can be said to establish a framework for a 'landscape of movements' or 'landscape of experience' – which is a way of conceiving/viewing the experience of the artwork as an entity consisting of the sum of movements, expressions and impressions through time and space. This landscape of possible relationships could be the actual definition of the artwork and then the 'landscape of possibilities' is the design tool contained in all its complexity in the 'extra–human'. The process is to define what kind of behavioural geography you want a tool for and then construct a behavioural entity to make such a tool.

'Smiles in Motion' – a concrete use of this design strategy

The creation of the installation 'Smiles in Motion' in 1999–2000 would not have been possible without an extensive use of these design methodologies. In this installation two people are connected through telematic technology. It is a machinery for augmented relationships using several means of telesensing and teleaction simultaneously, in a carefully designed and adjusted combination.

'Smiles in Motion' is an interactive set of furniture designed for augmented relationships between two people. Two chairs link two visitors, enabling them to converse with each other in a very special manner. This construction might be called a 'relation apparatus' and is able to transform speech into movement. Speech and sounds produced in the audible spectrum by the two visitors are converted into vibrations, through motors placed in the seats of the chairs. As a visitor perceives what is spoken in the form of vibrations, he is also shown the mouth of the other visitor on a monitor fixed in a globe. The visitors 'hear' each other through vibrations, synchronised with the images of the movements of their mouths and so may converse through vibrations and smiles.

Figure 1: 'Smiles in Motion' 2000 – www.boxiganga.dk © Kjell Yngve Petersen

In this artwork, the selection of telematic components, the decisions of how to process the media and meta–data streams and how to design the interface was a process of building the hardware, software and physical design around a series of test set–up with two performers being the advanced 'somebody' tool. We were able to adjust and combine on any number of parameters for extended periods of time, while having continuous and reliable feedback on the actual experience. It enabled us to develop the software that analysed their speech and converted it into motor movement while having real user feedback, not only on the actual sensing of vibrations coming from the chairs but also on the quality of the experience of the augmented relationship between two people using the chairs to converse.

In this project it made good sense and was very useful to put human beings at the centre of the process, not the least because, in the end, it was human beings who were going to receive the resulting artwork. We were not interested in simple measurable parameters that could be monitored by ordinary technology and by simple testing. We were creating an installation that was supposed to act through a complex combination of simultaneous senso–motoric relationships and our interest was more in how it supported a transformation of the quality of a conversation than in what was going on in a mere physical and technical way.

Bibliography

Ascott, Roy. (2003). *The Telematic Embrace, Visionary Theories of Art, Technology, and Consciousness*. Berkeley: University of California Press.

Barba, Eugenio and Savarese, Nicola. (1991). *The Secret Art of the Performer: A Dictionary of Theatre Anthropology*. London: Routledge.

Brandt, Per Aage. (2002). *Det Menneskeligt Virkelige (The Human Real)*. Copenhagen: PolitiskRevy.

Zeami:
The Zeami book is originally a diary written by Zeami Motokiyo (1363–1443), found in a bookstore in Tokyo in 1908, published in selection in japanese in the 1940s, translated to french in the UNESCO series: Connaissance de l'Orient , Zeami (1960) *La tradition secrète du Nô* , Traduction et commentaires de René Sieffert, Editions Gallimard, Paris. It was translated into danish in the Odin Teatret series: Teatrets Teori og Teknikk nr. 16 –.Zeami. (1971). *Den hemmelige tradition I No (The Secret Tradition of No)*. Denmark. Odin Teatrets Forlag. Translated from the french 1960 edition by Martin Berg. The English translation by J.Thomas Rimer and Yamazaki Masakazu is: Zeami (1984) *On the Art of the Nô Drama, The Major Treatises of Zeami*. Princeton University Press.

2.4 Electronic Cruelty

Gordana Novakovic

The world is divided into a complex caste system defined in direct proportion to the level of technological development. The flux of financial and information data exchange within a network of interconnected cities forms the Global City: the a–locus of superpower. The transparent hyper–real world of the obsolete horizon shaped by new technologies defines the contemporary aesthetics of abstraction and obsolete bodies. The citizen of the Global City is bombarded with the obscene pornographic banality of the mass media spectacle. Perception fractures and disperses, suffocating in noise. The body and mind are permanently overwhelmed with a kaleidoscope of noise: street noise, media noise, electromagnetic noise, genetic noise.

Immersed in the borderless ocean of the city, the contemporary citizen has the confidence that technological development has harnessed natural forces. Nature is a trophy, an ornament, an abstraction. The archaic fear of natural forces is replaced by the fear of technology and eternal progress. The force that sustains us is also that which destroys us. The network that forms the blood–circulating system of the Global City is spreading fear like a virus. Lulled by noise, bewitched by the spectacle of fear, in the screen–luminescent eternal twilight, global citizens are daydreaming artificial daydreams.

Interactive aesthetics has arisen in these conditions, and unless it strikes out its own path, it is in danger of turning into another form of tech–spectacle. Interactive installation, with its paradox of simultaneous repulsion and fear from the impersonal automatised process on the one hand and the acceptance of the oneiric immersion on the other, is the ultimate battleground between art and science and between the living body and technology. The symbolic conflict between man and machine takes on a ritual form. Instead of attempting to implant or reconstruct primordial ritual embedded in tribal society, interactive installation can be seen as a symbolic act of resolving contemporary tensions. Parallels are dangerous but useful, both where they fail and where they succeed. Consequently, I do not attempt to equate interactive installation with ritual but merely to use certain parallels that can shed light onto some specifics of interactive installation.

The definition of interactivity and spectacularity significantly changes with the hypothesis that interactive installation can take a ritual form. The root of the word spectacle is in Latin *specatculum* or *spectare*: to watch. It is related to the art of theatre that originated in and gradually replaced the ancient rituals in western culture. It refers nowadays to the blend of mass media and the entertainment industry, reflected in all segments of contemporary life to the extent that it has become a paradigm for contemporary social relations. However, it can be applied to a certain extent to ritual and interactive installation. This term is in opposition with the essence of both: the active participation of the audience is the *conditio sine qua non*, either in ritual or in interactive installation.

Spectacularity in interactive installation is of an entirely different nature than mass–spectacles. It is the fluid, changeable form of interactive installation that separates it from and opposes it to the uniform immersive anaesthetic of tech–spectacle. From the screen and virtual space of a particular personal computer, through endless spatial and dimensional diversities, interactive work merges virtual and real space each time in a unique manner. Custom–designed software and hardware architecture forms the basis of the contrast with typified entertainment industry production. The technology employed is to a large extent conspicuous as a constituent of the aesthetic. Non–linearity of segmented and unstable modules, consisting of loops in permanent change, is entirely circumstantial: intervals of participation replace continuous duration. It is the fusion of participant and technology in interaction that defines it and brings it into existence. In interactive work the process of interaction materialized in electro–magnetic and sound waves as a different class of matter is replacing the object of art. Einstein's theory of relativity, Heisenberg's principle of uncertainty, quantum physics –brought up a new aesthetics. 'Electronic ritual' can be embodied in an invisible flux –the blend of the installation's aura and participants' auras in a reverberating process of a cathartic collective experience within space/time/matter. Interactive installation as 'electronic ritual' might change the notion of seeing only through our visual sensory apparatus into an awakening of archaic 'seeing' as a complex cerebro–emotional process of perceiving the invisible: the participant 'sees feelingly'.

Participation in ritual is complete mental and physical engagement located in a very particular space. It is a closed event of the sacral genre that excludes spectators and audience. Only individuals who undergo a process of initiation are invited to participate. Physical activity in its endless varieties is inseparable from mental processes of total unity with space, resulting in transitions through levels of changed consciousnesses. Ritual operates in liminal spheres that are defined as a sensory threshold of changed consciousness introducing participants into esoteric meta–physicality. Participants reach deep levels of physical self–consciousness by dissolving their bodies in space and the symbolic dramaturgy. The entrancing experience leads to archetypal knowledge, reaching, according to Roy Ascott, even the primordial cell levels. Spectacularity in ritual functions is a major formal element. It creates a dramatic tension that symbolically signifies its inviolability and by designating the distinctness of the event separates it from the perception of everyday reality. Instead of voyeurisitic spectating there is active participation. It is the participant who actively creates dramatic tension through interaction and unity with the otherness of the event. Its metaphoric language personifies formidable forces beyond human control. There is an embrace and overcoming of primordial fears through uncanny, fearsome experiences.

Interactive work and ritual create drama through a language of signs and symbols in contrast to the logic of narration. The emblematic sonic and visual language of interactive installation and the specific radiant energy generated by its body, the reversible stream between the participants and environment through interaction, amalgamates participants' bodies with the installation parallel to the unity of body and space in ritual. Repetitiveness of the visual and the aural elements within a changeable flow of audiovisual modules as the common structure of interactive works operates as a classical mantric, trance–inducing ritual instrument. The sum of sensational stimuli changes the perception of time, space and matter leading to mental and physical self–awareness. The process of interaction transforms the characteristics and the apprehension of the particular space, incorporating and transfiguring technology.

The omnipresent conflict between man and technology is played out through the tension between the living body and the body of the installation. The human body becomes fluid, transparent, immersed and dissolved. Skin becomes a propulsive membrane. The sum of various sensations increases sensitivity and the level of self–consciousness of the body through a symbolic process of decomposing and recomposing. The participant's body is immersed in the environment; it feels and processes these impulses in its own right, reading the received data within but also beyond the levels of conscious perception. It is exposed not only to various audiovisual sensations but also to the installation's body generating different electromagnetic phenomena. However, a small number of works deliberately instrumentalize these effects. The way that interactive installations engage our sensory apparatus and the impact of the installation's environment on participants' personal bio–electric system is still enigmatic.

The aesthetics and functional mode of an interactive installation are significantly determined through the architecture of hardware and software. Regardless of their scale and complexity there is a division between works that can be called 'interactive instruments' and so–called 'responsive environments'. The structure of the 'interactive instrument' invites participants to follow a specified routine in order to establish interaction. Or they can be led by the 'shamanic' individuality of Stelarc, for example, whose body is in the role of mediator in interaction. On the other hand, there are so–called 'responsive environments'. Through a sensory system, the installation 'feels' and 'responds' to the presence of participants. A particular reaction that can be invoked by a responsive environment is the specific web of participants' trajectories through space and/or spontaneous gestures, a specific 'choreography' as a form of ritual activity. They can be immersed into an oneiric environment of intimate nature that involves the individual in a meditative trancelike experience, as in the works by Paul Sermon; or they can participate in Rafael Lozano Hemmer's spectacular phantasmagoric 'theatre of shadows' in an open public space. My installation works 'behave' as autonomous entities or as another participant in interaction. They should engage participants in a spontaneous dialogue of non–verbal communication mediated by non–tactile technology.

With interactive installation, the artwork is a disturbing autonomous entity, generating itself through an unstable process. The anxiety of entering dramatically charged dark spaces with unpredictable scenery is similar to the fear of entering phantasmagoric spheres of the unconscious. The instant feeling of unease conjoined with fascination releases primordial fears. Art can have a purifying function in overcoming fear of fear through the uncanny pleasure of experiencing it in controlled circumstances. We find examples in ancient cultures' rituals and art works, in medieval scenes of the Last Judgement and martyrdom, from Eleusinian mysteries, through Dürer's Apocalypse and Goya's phantasmagorias to various interventions on the human body in contemporary art. The metaphoric language of interactive installation is a powerful instrument that might create conditions for a cathartic experience of embracing and overcoming fears procreated by the global spectacle of fear and alienation caused by technology. The ritual nature of the cathartic collective experience of multiplying consciousnesses could be used as a basis for releasing a spirituality of a different class in interactive art that could oppose the materialistic terrorizing abuse of technology in global spectacle. It can bring about a form through which ritual can re–incarnate.

Bibliography

Ascott R. (1999). *Seeing Double. Art and Technology of Transcendence – Reframing Consciousness*. Exeter: Intellect.

Artaud A. (1961). *Collected Works*, vol. 2. London: John Calder.

Artaud A. (1964). *First Manifesto; Theatre and the Plague; Alchemist Theatre from Theatre and Its Double in Collected Works 4*. London: John Calder.

Baudrillard, J. (2003). *The Violence of the Global*, CTHEORY, vol. 26, 'La Violence du Mondial', in J. Baudrillard, *Power Inferno*. Paris: Galilee, pp. 63–83.

Capra, F. (1996). *The Web Of Life*. London: Harper Collins.

Capra, F. (1982). *The Turning Point*. London: Harper Collins.

Craig, G. (1999). *Theatre Past: The Arrival in On Theatre*) (ed. J. Michael Walton). London: Methuen Publishing.

Gennep, A. van. (1960). *The Rites of Passage*. London: Routledge and Kegan Paul.

Gombrowicz, W. (1960). *Pornografia*. New York: Grove Press.

Hawking, S. (1988). *A Brief History of Time*. New York: Bantam Doubleday, Dell Publishing.

Juenger, E. (1957). *The Glass Bees*. New York: New York Review of Books.

Juenger E. (1957). 'The Retreat into the Forest'. *Confluence*, vol. 3, no. 2 Harvard University

O'Hanlon, M. (1989). *Reading the Skin.* London: British Museum Publications.

Schechner, R. (1995). *The Future of the Ritual*. London: Routledge.

Turner, V. (1967). *The Forest of Symbols.* Ithaca and London: Cornel University Press.

Virilio, P. (1997). *Open Sky*. London: Verso.

Virilio, P. (2000). *Information Bomb*. London: Verso.

2.5 Design Against Nature

Anthony Crabbe

Disinfecting pits technology against nature. This issue is considered through examination of the design and development of a microwave disinfecting system for contact lenses. Here, technology intervenes to remedy a naturally occurring deficiency in human sight and the solution required by medical device accreditation requires a systematic annihilation of 'lesser' forms of life. The overarching question, then, is that of designer responsibility, beneath which lie beliefs and values that are seldom compatible. At first sight, these issues may appear to be merely peripheral to the design rationale of a disinfecting system. Closer inspection shows that simply to view these issues as contingent is in itself to adopt a particular view of responsibility.

The relatively recent discovery of invisible microscopic species has not evoked the kind of human sentiments evident in the discovery of other species, such as deep marine life. The implication of microbial life in human disease seems to have focussed more of our creative energy on their containment, rather than their development, which might be better warranted in view of their essential contribution to our own elementary biological processes.

The microbial problem in contact lens wearing is that the most common forms of lenses are only suitable to be worn during waking hours. Every evening, the patient is obliged to remove the lenses from the eye and store them in a container before reinserting them the next morning. This provides opportunity for the patient to introduce hazardous levels of micro–organisms onto the lens, which can cause painful, tenacious, sight threatening diseases, such as keratitis.

The industry has always recognised this limitation in daily wear lenses and has invested heavily in developing lens systems that obviate this problem. The two great hopes have been the daily disposable lens and the extended wear lens, which the patient only needs to remove and replace once a month. At present, both types of lens are suitable for about 70 per cent of contact lens prescriptions. The daily disposable is predictably more expensive than monthly or annual disposables, whereas the efficacy of extended wear lens for ocular health continues to be the subject of clinical debate (Weissman and Mondino 2003).

Contact lens wearing continues, then, to be a more expensive remedy than spectacle wearing and the cheapest, most widely used format is the daily wear lens. In 1996 a pharmacist friend and I, a product designer, became interested in designing a better disinfecting system for daily wear lenses. We believed one answer lay in making it possible for patients to irradiate daily their lenses, solution and storage case in their domestic microwave ovens. Our main motivations in undertaking the kind of design and development that would normally be done by major pharmaceutical companies were to create a lens care system that was both safer and cheaper for patients. We saw that the care industry was dominated by an oligopoly of four major corporations, including Nestlé and Ciba Gigy. Along with others, we thought their prices were more justified by their market position than their R&D costs (Monopolies and Mergers Commission 1993) and that the disinfecting efficacy of their products could be better (Lowe et al. 1992).

Microwave treatment offers complete rather than partial kill of all challenge organisms. It requires only a basic saline solution, free of preservatives and additives used to justify relatively high prices, yet associated with allergic reactions (Herbst and Maibach 1991). Furthermore, the most common 'multipurpose' solutions seemed to encourage poor hygiene compliance, with patients ignorant of the fact that the disinfecting effect was greatly diminished if they failed to rub and rinse their lenses and storage cases thoroughly (Kilvington 1993)

The alternative we designed and developed was a plastic unit for treating lenses in a microwave and a modified solution appropriate for irradiation treatment (see fig. 1). Simple as this sounds, it committed us to a gruelling EU medical device accreditation process lasting more than two years, in which we had to prove beyond reasonable doubt that the system was safe to advocate and market to patients and did not damage the many different types of soft lenses they used. The process is very expensive. It required us to raise grants and, later, venture funding to set up an aseptic production plant and quality systems for manufacture and distribution.

Figure 1: Disinfector Unit

Rather than rehearse here in detail the various challenges of this undertaking, I want to return to the other less obvious design responsibility issues that underlay our chosen path. The design problem and the form of the solution were largely predetermined by the design of daily wear lenses. Consumer protection law required us to prove specific rates of kill for standard 'challenge' micro–organisms. So just by choosing the design problem we were endorsing an environmental strategy of eradicating all microbial life from the lens storage area.

In this respect, ours was just one more of a growing number of products adopting this strategy to protect the consumer from the natural hazards of their personal environment. It is not only the consumer's foodstuffs and cosmetics that are defended by disinfecting treatments and preservatives but entire rooms, such as the kitchen and bathroom, for which there are ever more aggressive bactericidal cleaning agents.

Since microbial life is not sentient, it might appear rather ridiculous to question the responsibility of designing better ways of killing microbes. However, the question is a fair one in consideration of the problems generated by aggressive and indiscriminate disinfecting, a strategy which might be termed 'bio–massacre'. Even the general public is aware of the phenomenon of 'super bugs', micro–organisms that have developed resistance to common drug treatments. Some have argued that just as indiscriminate use of antibiotics promotes the development of resistant strains of micro–organisms, blanket use of disinfecting agents can do the same thing by creating competition–free environments conducive to the evolution of resistant mutant strains (Levy 2001; McMurry et al. 1998). The press scares about 'super rats' (rats with an acquired immunity to the poison Warfarin) testify that this mutation phenomenon also applies at the level of sentient life forms (Daily Telegraph 2000).

There are, then, purely strategic reasons for questioning the bio–massacre approach and the choice of such an emotive label is intended to alert others to the potentially dangerous consequences. Some may say that there is no room for emotional advocacy in the sciences but the present debates over the efficacy of biotechnological developments such as GM foods and cloning show this to be a wish, not a fact. Science is the product of emotional beings and operates within a cultural context. In this culture, commerce provides the necessary support to transform scientific concepts into technologies. In turn, commerce relies on marketing to cultivate consumer desire and the emotional impact of 'germ threats' to consumers is a proven market strategy for creating new product demand.

Science is also playing an active part in transforming our theological visions of the world in terms of adjusting both our sense of universal scale and our understanding of just how special is the presence of life on our speck of a planet. Our understanding that the progenitors of this statistical miracle were the simple species our technology is attacking so indiscriminately seems bound to invoke non–scientific reasoning, particularly about our own role within the natural scheme. Speculation of this sort appears to reveal two rather different views of what may called 'natural determinism'.

On the one hand, we may view nature as a wonderfully complex and finely tuned system, now under threat from our technological hubris, which unchecked, will doom many species, including our own, to a tragic fate. On the other, it may be argued that technology is simply a product of this system and the dire consequences for other species is only a matter of the natural system reconfiguring itself to favour the most evolved and sophisticated of its species. Portrayed in theological terms, the first view could be characterised as a kind of mystery faith, akin to the animism of native Americans, the Buddhist sanctification of all life forms or the divination of primal forces like Gaia (Lovelock 1979). The second view is anthropocentric, more akin to an Old Testament vision, where God, the highest authority, has made this world in order to accommodate his greatest creation, to whom he has granted dominion over all its other creatures.

The mystery faith issues dark prophecies to advise technological caution, whilst the anthropic faith evangelically promotes progress and growth (especially of markets). Like it or not, most designers are servants of the anthropic faith, which they may seek to redirect from within, by espousal of 'environmentally friendly' strategies, but over which they have little real control, lacking the power or influence of a government that sets the agenda for consumer protection regulation and can choose to ignore initiatives to globalise environmental strategies.

The most powerful (and obdurate) of these governments includes among its constituents many of the world's leading developers of contact lens research and marketing. Few beneficiaries of their products may be inclined to renounce them on the basis of compassion for the microbial species that threaten to subvert product efficacy. Contact lens related diseases are painful and sight threatening. Yet the literature shows the most serious diseases are relatively rare (Nilsson et al. 1999; Radford et al. 2002) and can often be attributed to deficiencies of human nature, especially ignorance and laziness (Claydon et al. 1994).

Hence any endeavour to design better lens wearing regimens pits the designer against both the natural world and human nature. Having committed ourselves to the bio–massacre strategy, we believed a major advantage of our system was that it made it much harder for the patient not to comply with the clinical necessity of disinfecting lenses, solution and all parts of the storage case. In the event, the market reaction of our actual customers, optometrists, told us that our design was too inconvenient in requiring a change of care regimen from the bathroom to the kitchen. Furthermore, many optometrists were reluctant to take a product that in claiming 100 per cent disinfection invited unwelcome questions about the percentage kill achieved by their established products!

Late in the day we did modify the design to one that used a multipurpose solution and a much smaller lens case that gave patients the option to choose either a maximum effect 'hot' disinfecting treatment in the kitchen, or should that be inconvenient, a standard 'cold' treatment in another site (see fig. 2). Unfortunately

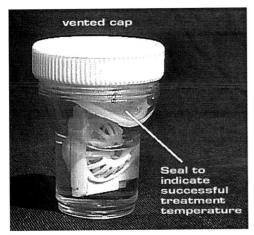

Figure : Mark II

the terms of our venture funding gave us insufficient time to make this adjustment, committing us to a single strike with our pilot design.

In private post mortems of our approach, a few leading clinicians raised doubts about the benefits of trying to improve the daily wear regimen. They saw such an endeavour as a stopgap measure whilst the industry as a whole was delivering the best design solution by developing daily disposable and extended wear lenses. In view of the design responsibility issues raised above, this indeed seems a sensible view. Both systems effectively bypass the need for bulky energy consuming care products that wage indiscriminate war on naturally occurring micro–organisms.

On the other hand, recent market figures suggest that the optimisation of the new designs is still some way off, since daily wear lenses continue to hold their ground and dominate market share (ACLM 2002). In this sense, the stopgap period is a remarkably long one, exceeding the expected life cycle for many other consumer products. Apart from the time it takes to move the dispensing profession from familiarity with one type of system to another, a key factor for the survival of daily lenses seems to be price. As long as daily disposables and extended wear lenses can command a premium price in respect of their novelty and daily wear can cut costs to preserve their market share, the benefits of newer designs will be harder to pass on to patients.

During this hiatus the industry continues to hedge its bets by carrying on with new research and development in daily wear care products. The most notable recent development has been the 'hands off' chemical systems, which claim to obviate the need for the patient to rub and rinse the lens and storage case. These developments reinforce the sense of how design, research and development are merely aspects of a much more complex system of free–market technology in which the designer's best solutions are not necessarily the most effective product solutions.

The challenge facing designers, then, is not simply a choice between two views of responsibility. It also involves facing a third and perhaps more powerful 'theology' –that of mammon. Even its most ardent devotees would probably accept that the relationship between retailer and consumer is not rational in the objective scientific sense. We discovered first hand that 'convenient' may be a better marketing pitch than 'safer' or that 'complete' disinfection is potentially more of a market threat than a vague promise of 'disinfection' (actually meaning a reduction in some but not all challenge organism populations). The marketing campaign is waged as much for customer hearts as minds and usually in deference to consumer convenience.

The challenges of design responsibility are then formidable. Irrespective of their backgrounds in sciences or humanities, designers need to acquire understanding of more than one discourse to stand any chance of satisfactorily meeting the challenges of responsibility. This appears a daunting task in an age of increasing specialisation and may explain the growing attraction of working in multidisciplinary rather than specialist project teams.

Bibliography

Association of Contact Lens Manufacturers. (2002). *Market Trends* (internal document).

Claydon, B.E. and Efron, N. (1994). 'Non Compliance in contact lens wear'. Ophthalmic Physiol Opt. 14 Oct. 1994, pp. 356–64.

Daily Telegraph. (2000). 'Super rats pose threat to Britain', 9 Nov.

Herbst, R.A. and Maibach, H.I. (1991). 'Contact dermatitis caused by allergy to ophthalmic drugs and contact lens solutions'. *Contact Dermatitis*. Nov. 25 (5), pp. 305–12.

Kilvington, S. (1993). 'Acanthamoeba trophozoite and cyst adherence to four types of soft contact lens and removal by cleaning agents'. *Eye.* 7 (Pt 4), pp. 535–8.

Levy, S.B. (2001). 'Antibacterial Household Products: Cause for Concern'. *Emerging Infectious Diseases* 7:3 Supplement.

Lovelock, J. (1979). *Gaia: A New Look at the Earth*. Oxford: Oxford University Press.

Lowe, R., Vallas, V. and Brennan, N. A. (1992). 'Comparative efficacy of contact lens disinfection solutions'. *CLAO Journal*, 18:1, pp. 34–40.

McMurry, L.M., Oethinger, M. and Levy, S.B. (1998). Triclosan targets lipid synthesis. *Nature*, 394:531–2.

Monopolies and Mergers Commission (1993). *Contact lens solutions: A report on the supply within the United Kingdom of contact lens solutions*. Summary of conclusions, HMSO, pp. 207–8.

Nilsson, S.E. and Montan, P.G. (1994). 'The annualised incidence of contact lens induced keratitis in Sweden and its relation to lens type and wear schedule: results of a 3 month prospective study'. *CLAO J.* 20 4, pp. 225–30.

Radford, C.F., Minassian, D.C., Dart, J.K. (2002). 'Acanthamoeba keratitis in England and Wales: incidence, outcome, and risk factors'. *British Journal Opthalmology*. May; 86(5), pp. 536–42.

Weissman, B., Mondino, B.J. (2003). Why daily wear is still better than extended wear. *Eye Contact Lens*. Jan. 29 (1 Suppl):S145–6; discussion S166, S192–4.

2.6 Why Look at Artificial Animals?

Geoff Cox and Adrian Ward

> *Animals are not machines ... Actually only machines are machines. Nothing else is made by human beings from parts and for purposes entirely supplied by themselves. Nothing else therefore can be understood simply by reading off those parts and purposes from the specifications*
>
> (Midgley 1979 p. xvi)

> *But basically machines were not self–moving, self–designing, autonomous ... Now we are not so sure ... Our machines are disturbingly lively, and we ourselves frighteningly inert.*
>
> (Haraway 1991 p. 194)

Animals are both like and unlike humans. If this were partly reinforced by human isolation from the wider world of nature under the culture of capitalism under late techno–capitalism, animals could be said to be increasingly both like and unlike machines –or to put it another way, machines are increasingly being classified according to the model of the animal. The interrelationships are enduring ones, reactivated by changes in social and technological production, making the former distinction further complicated by the addition of artificial life–forms and the merging of biological and computational forms. The task of classifying and differentiating between animals, humans and machines is one performed with increasing amounts of difficulty, born out of complexity, to use an operative term.

In his essay 'Why Look at Animals?' (Berger 1980), John Berger states: 'They belonged there and here. Likewise they were mortal and immortal. They were subjected and worshipped, bred and sacrificed' (Berger 1980 p. 5). Pointing to the use of the connective 'and' when 'but' might be more easily anticipated, Berger reveals the inherent dualism in our historical relationship with animals. The distinction expresses a contradictory (even dialectical) impulse bound up with human evolution –from lowly four–legged beasts used as metaphor to two–legged ones with the advanced ability to use metaphor:

> *If the first metaphor was animal, it was because the essential relationship between man and animal was metaphoric. Within that relation what the two terms –man and animal –shared in common revealed what differentiated them. And vice versa*
>
> (Berger 1980 p. 5)

According to Berger's thesis, animals begin to disappear during the process of urban industrialisation in the 19th century. In parallel, domestic pets multiplied and adhered to the truism of resembling their masters in that they were separated from their natural way of life and made to lead artificial lives in domestic environments. These pet artificial animals reflect the alienated conditions of late capitalism, expressing sentimental attachment and preferred property relations –but not the parallel autonomy of previous times. The pet and owner both lack agency (the power to act independently), as one jerks the other by its lead or responds to stimuli like Pavlov's dog. As part of the bourgeois home, the symbolic virtual animal also follows this sad trajectory, reduced to appearances (in the spectacle), encapsulated in the over–production of anthropomorphic materials designed for family viewing and excessive consumption (far too many examples to mention but Berger cites the work of Beatrix Potter and Disney). Children are the key players here, of course, as they are seen not only to like animals but also to be like animals and hence required to engage with them as part of a process of socialisation through reflexive play (feed them, train them, take them to bed, bury them and so on). This phenomenon is ever developing, ranging from the adoption of real and toy animals to the more recent realistic animal toy robots –the Sony Cyberdog AIBO and Tamogotchi come to mind. Slavoj Zizek's essay 'Is It Possible to Traverse the Fantasy in Cyberspace?' (1999) describes these virtual pets as instruments of 'interpassivity' (rather than interactivity), turning children from carers into virtual murderers. Clearly, ideology is at work in the ways we observe animals and choose to characterise our relation to them in the home and in more public spaces.

Similar anthropomorphic activity is often found when dealing with the computer. Most people have talked to their computer at some point; many assume that malfunctioning software is the result of bad behaviour and that good behaviour can be encouraged by treating the machine gently or according to behaviourist principles. Operating systems are guilty of propagating these myths too: you can get your Macintosh to say 'It's not my fault!' whenever a crash occurs. The *Vivaria.net* project contains software artworks that approach (life–like) code in a quite different manner, challenging the usual representational forms and questioning the ways in which programmers/artists look anthropomorphically at their code/artwork –seeing it create, perform, generate unexpected results, interact with other objects, feed, reproduce, crash/die and so on.

Making reference to the Tamogotchi, *Carbon Lifeform* software suggests that it is not necessary to artificially enhance the life–like appearance of something that purports to be a pet. Most pet software visually simulates an animal of sorts, lending a bitmap body to an entity that exists only as code, data and metadata. *Carbon Lifeform* consists of standard GUI widgets that are used to convey the general status of the pet. Its body is a UI–compliant document window, with a close and minimize button. The only suggestion of a simulated animal body is through the use of the happy/sad Mac icon that is used by classic Macintoshes to alert the user to hardware problems. Once installed, *Carbon Lifeform* needs to eat regularly –it takes bytes out of your files (so don't feed it anything important) and requires regular attention. Should your *Carbon Lifeform* die, your computer will die too (by shutting down without warning). Keeping the *Carbon Lifeform* alive is therefore a rather more critical task –and one that revolves around primitive logic routines rather than complex behavioural patterns often deployed by A–life softwares. Perhaps *Carbon Lifeform* helps to regain a sense of autonomy that animals have lost in the last two centuries, if only because it does not conform to the same bourgeois ideas of what constitutes being a pet or artificial animal.

Rather like the objects in a museum, there exists a central paradox at work in the destruction of the natural world with its simultaneous preservation. In the case of zoos, Berger is saying that freedom as well as visibility is made artificial. We observe living objects as if they were dead. The animal inhabits an artificial natural world, a simulated or virtual world of rocks and trees as fake as theatrical props. The animals are isolated and do not interact with other species and they become as dependent as pets on their keepers for food and social arrangements and interactions, including the supply of mates for reproduction. In other words, zoos are mausoleums to life and survival and monuments to historic loss. Berger says: 'The zoo to which people go to meet animals, to observe them, to see them, is, in fact, a monument to the impossibility of such encounters' (Berger 1980 p. 19). In the zoo, captive animals perform a symbolic but passive function to endorse scientific, economic and colonial power. It has become commonplace for artists to use biological metaphors and examine creativity in the light of scientific investigations in artificial life, simulating the characteristic processes of living things but what values are being reproduced alongside these works? It is now obvious that animal and machine (or organic and technical) processes are analogous and similarly contain self–organising functions but are there corresponding over–simplifications and an uncritical affirmative tendency at work?

Berger draws the comparison with art: 'In principle, each cage is a frame round the animal inside it. Visitors visit the zoo to look at animals. They proceed from cage to cage, not unlike visitors in an art gallery who stop in front of one painting, and then move on to the next or the one after next. Yet in the zoo the view is always wrong' (Berger 1980 p. 21). Perhaps this line of thinking serves to suggest that we now look at animals wrongly in new ways. According to Nichols (in 'The Work of Art in the Age of Cybernetic Systems'), the zoo exhibits the logic of a self–regulating system and simulated animal–nature and natural environment (Nichols 1988 p. 34) –much like the idea of virtual worlds presumes the (real) world as we perceive it to be real. Nichols captures the debate about artificial life through Benjamin's artwork essay on reproduction in questioning the presumptions that are made about what constitutes art or life.

Another example from the *Vivaria.net* project is *animal.pl*, a single piece of software (art) that runs continuously on a server connected to the Internet. It performs many different activities, each of which is carefully prescribed and coordinated by the author to interact with other parts of the same software. It is through this self-sufficient design that the analogy of a life-form is drawn. *Animal.pl* has an agenda, a voice, a policy, a purpose and even a lifeline. All this without a single line of A.I. or A-life code. The software is written not from a scientific perspective (which merely propagates the equally restrictive notion that software is purely functional) but from a personal one –that of the author –where emotion and feeling is expressed through the craft of coding and not through a mere simulation of feeling. Even so, *animal.pl* requires a connection to the Internet to operate and requires other (human) animals to contribute to it before it can become active. In this sense, it externalises its complexity through its interfaces, demonstrating life-like behaviour. The software may even evolve into a new species as a result of its own actions –the last prescribed action *animal.pl* performs is to apply the GNU Public License to itself, offering itself for others to modify and adapt in new ways.

Undoubtedly former firm distinctions between animals, machines and humans are now unreliable –though, of course, the idea of the zoo was partly to reinforce the distinction in the light of Darwinism. On the contrary, life can now be generated by what Turkle calls 'unnatural selection' (Turkle 1997 p. 149) –which seems even more unsettling than the former premise. How would one begin to establish whether life has been demonstrated or produce some kind of taxonomy based on this?

The *Vivaria.net* project attempts to classify artificial life-forms by appropriating Darwin's 'Divergence of Taxa' diagram and, at least provisionally, using the four characteristics of life suggested by the Los Alamos conference on A-life (in 1987). The dynamic taxonomy allows artists and programmers to add their projects to the diagram and identify the behavioural characteristics of the code –rather like a software repository. Furthermore, the aim is to express the idea of the diagram as a kind of organism in its own right. The diagram will evolve in a way that might express conflicting criteria and a spurious scientific methodology. Users may add new branches to the diagram, as well as offer links to artificial life-forms that should be classified under a particular branch. Are we simply suggesting that we categorise what constitutes life wrongly in new ways too?

You can trace the historical lineage here from artificial intelligence, as well as the influence of chaos theory in believing that mathematical structure lies beneath apparent randomness and that randomness could generate mathematical structures. However, unlike traditional A.I. thinking, A–life relies on its fundamental equivalence to real life. Hence it is deeply controversial and ideologically–charged. When looking at artificial life, it appears tamed by the computer screen in much the same way as watching wild animals in a zoo; you are separated by the cage bars or glass. Furthermore, if once watching animals allowed humans to imagine being at one with nature, how does the human respond to the discovery that nature itself is programmable? In the culture of simulation, there is nothing natural about the way we look at these animals, artificial or not –we look at them artificially in new ways. Is the distinction between humans, animals and machines undermined or reinforced? Rather than the proliferation of animal representations being compensatory to historic loss, they have added to the disappearance according to Berger, rendered almost entirely distant by close inspection at a zoo or laboratory. If animals become ever more exotic and remote (Berger 1980 p. 24), is the attraction of artificial life merely its contemporary form? Berger thought the dualism between animals and humans had been lost and went further to suggest this as a link to totalitarianism. He continues bleakly: 'This historic loss, to which zoos are a monument, is now irredeemable for the culture of capitalism' (Berger 1980 p. 26). Berger made that statement in 1980. Is this loss even more pronounced under present technological and cultural conditions? In looking, the imperative must be to shift not just from a politics of representation (the critical orthodoxy at that time) but a politics that takes additional account of generative processes and questions of autonomy.

Why look? It would appear that under the conditions of techno–capitalism, humans are both like and unlike artificial animals, which is why we insist on posing the question in the first place.

Bibliography

Berger, J. (1980). 'Why Look at Animals?' in *About Looking*. London: Writers and Readers.

Haraway, D. (1991). *Simians, Cyborgs and Women: The Reinvention of Nature*. London: Free Association.

Midgley, M. (1979). *Beast and Man: The Roots of Human Nature*. London: Methuen.

Nichols, B. (1988). 'The Work of Culture in the Age of Cybernetic Systems', in *Screen,* vol. 29, no.2, Winter. pp. 22–46.

Turkle, S. (1997). *Life on the Screen: Identity in the Age of the Internet*. London: Phoenix.

Vivaria.net', http://www.vivaria.net/, STAR in collaboration with Signwave.

Zizek, S. (1999). 'Is It Possible to Traverse the Fantasy in Cyberspace?', in Elizabeth Wright and Edmond Wright, *The Zizek Reader*. Oxford: Blackwell, pp. 102–24.

2.7 Biopoetry

Eduardo Kac

Since the 1980s poetry has effectively moved away from the printed page. From the early days of the minitel to the personal computer as a writing and reading environment, we have witnessed the development of new poetic languages. Video, holography, programming and the Web have further expanded the possibilities and the reach of this new poetry. Now, in a world of clones, chimeras, and transgenic creatures, it is time to consider new directions for poetry *in vivo*. Below I propose the use of biotechnology and living organisms as a new realm of verbal creation.

1. Microbot performance: Write and perform with a micro–robot in the language of the bees, for a bee audience, in a semi–functional, semi–fictional dance.

2. Atomic writing: Position atoms precisely and create molecules to spell words. Give these molecular words expression in plants and let them grow new words through mutation. Observe and smell the molecular grammatology of the resulting flowers.

3. Marine mammal dialogical interaction: Compose sound text by manipulating recorded parameters of pitch and frequency of dolphin communication, for a dolphin audience. Observe how a whale audience responds and vice versa.

4. Transgenic poetry: Synthesize DNA according to invented codes to write words and sentences using combinations of nucleotides. Incorporate these DNA words and sentences into the genome of living organisms, which then pass them on to their offspring, combining with words of other organisms. Through mutation, natural loss and exchange of DNA material new words and sentences will emerge. Read the transpoem back via DNA sequencing.

5. Amoebal scripting: Hand–write in a medium such as agar using amoebal colonies as the inscription substance and observe their growth, movement, and interaction until the text changes or disappears. Observe amoebal scripting at the microscopic and the macroscopic scales simultaneously.

6. Luciferase signalling: Create bard fireflies by manipulating the genes that code for bioluminescence, enabling them to use their light for whimsical (creative) displays, in addition to the standard natural uses (e.g. scaring off predators and attracting mates or smaller creatures to devour).

7. Dynamic biochromatic composition: Use the chromatic language of the squid

to create fantastic colourful displays that communicate ideas drawn from the squid Umwelt but suggesting other possible experiences.

8. Avian literature: Teach an African Grey parrot not simply to read and speak, and manipulate symbols, but to compose and perform literary pieces.

9. Bacterial poetics: Two identical colonies of bacteria share a petri dish. One colony has encoded in a plasmid a poem X, while the other has a poem Y. As they compete for the same resources, or share genetic material, perhaps one colony will outlive the other, perhaps new bacteria will emerge through horizontal poetic gene transfer.

10. Xenographics: Transplant a living text from one organism to another, and vice versa, so as to create an *in vivo* tattoo.

11. Tissue text: Culture tissue in the shape of word structures. Grow the tissue slowly until the word structures form an overall film and erase themselves.

12. Proteopoetics: Create a code that converts words into amino acids and produce with it a three–dimensional protein poem, thus completely bypassing the need to use a gene to encode the protein. Write the protein directly. Synthesize the protein poem. Model it in digital and non–digital media. Express it in living organisms.

13. Agroverbalia: Use an electron beam to write different words on the surface of seeds. Grow the plants and observe what words yield robust plants. Plant seeds in different meaningful arrays. Explore hybridization of meanings.

14. Nanopoetry: Assign meaning to quantum dots and nanospheres of different colours. Express them in living cells. Observe what dots and spheres move in what direction, and read the quantum and nanowords as they move through the internal three–dimensional structure of the cell. Reading is observation of vectorial trajectories within the cell. Meaning continuously changes, as certain quantum and nanowords are in the proximity of others, or move close or far away from others. The entire cell is the writing substrate, as a field of potential meaning.

15. Molecular semantics: Create molecular words by assigning phonetic meaning to individual atoms. With dip–pen nanolithography deliver molecules to an atomically flat gold surface to write a new text. The text is made of molecules which are themselves words.

16. Asyntactical carbogram: Create suggestive verbal nanoarchitectures only a few billionths of a metre in diameter.

17. Metabolic metaphors: Control the metabolism of some micro–organisms within a large population in thick media so that ephemeral words can be produced by their reaction to specific environmental conditions, such as exposure to light. Allow these living words to dissipate themselves naturally. The temporal duration of this process of dissipation should be controlled so as to be an intrinsic element of the meaning of the poem.

18. Scriptogenesis: Create an entirely new living organism, which has never existed before, by first assembling atoms into molecules through 'atomic writing' or 'molecular semantics'. Then, organize these molecules into a minimal but functional chromosome. Either synthesize a nucleus to contain this chromosome or introduce it into an existent nucleus. Do the same for the entire cell. Reading occurs through observation of the cytopoetological transformations of the scriptogenic chromosome throughout the processes of growth and reproduction of the unicellular organism.

Illustrations

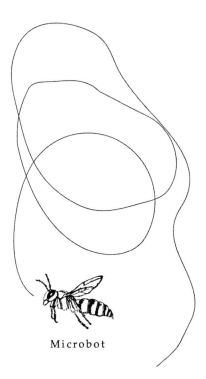

Microbot

Figure 1: The 'robeet' (robotic bee) would allow a poet to write a performative dance–text that has no reference in the physical world (that is, does not send bees in search of food). Instead, the new choreography (kinotation) would be (bee) its own reference.

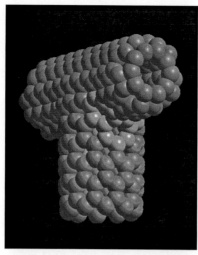

Figure 2: The beginning of a new alphabet. Letters can be created with carbon nanotubes, tiny cylinders only a few billionths of a metre in diameter, as exemplified by this letter 'T'. Words created at this nanoscale can be made stable under the laws of quantum molecular dynamics. The first letter of the word 'Tomorrow'.

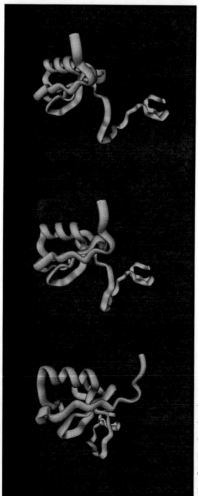

Figure 3: By assigning specific semantic values to amino acids, a poet can write a protein. The 'Genesis' protein, above, critically encodes the biblical statement: 'Let us make man ... have dominion over the fish of the sea, and over the birds of the air, and over every living thing that moves upon the earth' (Genesis 1: 26, 28).

Figure 4: Start by collecting mud from the bottom of a lake or river. Create a flat sealed box and introduce the mud, supplemented with water from the lake or river, cellulose (use the most interesting pages of a newspaper), sodium sulphate and calcium carbonate. Seal the box. Approximately 5000 different micro–organisms (prokaryotic bacteria and archaea) will make up this population. Make a mask with the text to be read. Expose to light everything but the text. In about two weeks the text will be dark enough to be clearly read. Expose the whole surface to environmental light and allow the words to dissipate. The population within the sealed chamber will recycle nutrients and will support itself with no additional aid

3. The Place
3.1 Real Virtuality: Authenticity in Electronic and Non-electronic Environments

Eril Baily

A perennial issue between so-called reality and virtual reality (or non-electronic and electronic environments) is the dubious authenticity of a virtual environment. As the qualifying adjective 'virtual' implies, virtual reality or electronic environments are considered:

1) Almost nearly but not quite reality. For instance, we often say, something can be considered the case and the fact that it is not quite the case can be overlooked. In this instance, virtual reality is granted the status of reality; it will 'pass' for reality because the 'lack' or deficit is too small to bother with. Hence, to all intents and purposes it is as authentic as reality even though the environment might be at variance with reality –indeed, might be quite fantastic. Immersion in a virtual environment often elicits this evaluation –a participant exclaims, 'God, it was virtually reality.' The participant recognises in hindsight that they lost themselves within the illusionary, illusory world and their experiences were just as comprehensive and committed as those in reality. This is a positive response –it acknowledges the power and 'magical' veracity of the virtual. It is an exclamation of wonder.

But, it is important to note, however, the virtual is, in the final analysis, not mistaken for reality. Should the participant not, at some point, acknowledge the phantasmagoric aspect of the virtual environment he would be considered insane because unable to distinguish between reality and phantasy[1]. In other words, in our current context the ability to discriminate between reality and phantasy –to wake up from the dream, as it were, and to recognise the dream was a dream –demarcates the psychically functional subject from the psychically dysfunctional one. And, in some sense, we consider those psychically dysfunctional subjects to have 'lost' or not 'developed' a crucial aspect of consciousness –the ability to distinguish between authentic and inauthentic environments; between reality and the imaginary.

2) Virtual reality has the effect and affect of reality but is not actually reality. This is the same case as above –but the lack or deficit –the fact that it is phantasy is taken into account within the experience. It is acknowledged that experience within the virtual can be equally as compelling as that of reality –but the participant does not 'lose' himself within the virtual environment. In fact, should he

do so, he would feel he had been duped –hoodwinked. To have lost oneself in something that was only an illusion attests to a lapse in discrimination –if not a loss of consciousness.

In this case virtual reality is not accorded the status of reality. It does not pass for reality because the focus of the evaluation is in precisely keeping the distinction between phantasy and reality in consciousness. This is a position taken by the critic –to comment on one's experiences as one undergoes the experience, to be conscious of self as an object of experience.

3) Virtual reality is an image of reality. In a contemporary context this can be considered an ordering or hierarchy of reality and image. The first order is the most 'common' and 'simple' meaning of the virtual. The virtual image is a representation of reality –a replication, reflection or a copy. In this instance reality and its image are cross–referenced and the image evaluated in terms of its isomorphism with reality. The status of the image is in proportion to its fidelity to reality. In this case, there is nothing fantastic in this virtual reality because reality, itself, excludes the fantastic. That is, one of the criteria of reality is that it is not constituted from imaginary, illusionary or 'out of this world' components.

At the second order (and on), that is, in the order of the simulacra, the image and its relation to reality is more complex. Reality, in the conventional, material sense of it, has been eclipsed by an image of it, which in turn has been represented by an image. That is, 'reality' is an image of reality that is replicated, reflected or reproduced. But, although this confounds what is 'taken' or accepted as reality, the second order image is cross–referenced to the first order image (as reality) and evaluated in the same way as it was in the more simple case. In other words, the status of the simulacra is in proportion to how faithfully it replicates an image of reality.

I have elaborated on these various senses of the term 'virtual' and what they imply or entail because I think there are three issues that are worth isolating and considering.

1) In each case reality is assumed to precede and pre–ordain that which is the virtual, or 'illusionary'. Furthermore, reality and virtuality are dichotomies in which the virtual is that which reality is not. That is, the virtual is all that which is excluded from defining reality and is positioned as other to reality. Consequently, virtuality is dependent upon reality for its coordinates as a locus and has no meaning other than the negative of reality.

2) To not discriminate between reality and virtuality is to have lost an essential

capacity of consciousness. To live in an illusionary environment means at the very least one's life lacks substance and *in extremis* renders one 'insane'. It is to live an unauthenticated life because no other can confirm the veracity of the habitat. Unlike an imaginary environment, which is solitary and private, reality is public. It is an accepted fact that more than one subject inhabits reality whereas it is an accepted fact that the insane person lives in a private world –he is the sole inhabitant of the phantasy.

3) To 'lose' consciousness of oneself in the illusionary is acceptable but only on condition one relocate oneself in reality. In reaffirming one's locus in reality, the subject may implicitly or explicitly acknowledge the experiential impact of the virtual and its capacity to virtually eclipse reality but at the same time makes clear the distinction between them. To say, 'What a vivid dream I had', simultaneously acknowledges the power of phantasy and that it *is* phantasy.

Should this distinction not be made, should one not be able to retrieve oneself from phantasy, signifies a loss of consciousness. To lose and not recover consciousness of oneself is a tragic loss of humanness –tragedy understood in the classic sense of a fatal flaw, a capacity for greatness that is thwarted by a vulnerability within the subject.

I now want consider these issues and try to give an account of some aspects of contemporary life that I believe are symptoms or indices of changes in consciousness and perforce signal changes in what it means to be a subject: to be human and what we take as reality.

I will appeal to psychoanalysis –particularly Freud (1856–1939) and Lacan (1901–1981) for ways of reconsidering the relations between virtuality and reality and some aspects of consciousness.[2]

Freud consistently claimed throughout his writings on his psychoanalytic practice and theories that arriving at a distinction between phantasy and reality was a major psychical undertaking and at best was never fully accomplished. According to Freud, infancy and childhood is a reckless period of free–playing phantasy. He writes, 'we should equate phantasy and reality…[and] not bother to begin with whether the childhood experiences under examination are the one or the other' (Freud 1916–17).

It is not until the little girl and little boy acknowledge the loss of the mother's phallic power that unconscious processes and content (phantasy) is demarcated from conscious process and content (reality). In the process, anti–social aims are repressed and socially sanctioned aims are consciously upheld and pursued.

Accomplishing this inaugurates unconscious and conscious dynamics and consolidates the constitution of the subject.

Nevertheless, the so-called reality principle (socially sanctioned aims and activities) is always in jeopardy of being usurped by the pleasure principle (phantasies of the satisfaction of taboo wishes and desires), in other words, dreams, daydreams, neuroses, hysteria, parapraxis etc.

The important point for my purposes is that this tenuous distinction between phantasy and reality is effected through phantasy —through an image that is so compelling that it must be repressed and thereby constituting the difference between unconscious and conscious processes. That is, in a doubling up of phantasy, the phantasy of the mother's phallic power is exposed as phantasy by the phantasy of her castration —at which point she is 'positioned' in reality. The dynamics of power as they operate in reality are exposed through the overwhelming authenticity of the phantasy —through the indubitable reality of the virtual. Thus, real virtuality constitutes the distinction between phantasy and reality which, in turn, once reality is constituted, positions phantasy as the repressed other relative to reality.

Indeed, this account of the potency of the virtuality would seem to destabilise the first consideration —namely that reality precedes and pre-ordains the relations of phantasy to reality.

I will now give a brief account of Lacan's mirror stage and the salient features I think worth considering.

Lacan argued that the impetus to the development of the ego and its concomitant locus in reality is the infant's captation by its mirror image. He writes:

> *The child, at an age when he is for a time, however short, outdone by the chimpanzee in instrumental intelligence, can nevertheless already recognize as such his own image in a mirror...This jubilant assumption of his specular image by the child at the infans stage [anywhere from six to 18 months] ...would seem to exhibit in an exemplary situation the symbolic matrix in which the I is precipitated in a primordial form ... The important point is that this form situates the agency of the ego, before its social determination, in a fictional direction . . . The mirror-image would seem to be the threshold of the visible world ... I am led, therefore, to regard the function of the mirror-stage as a particular case of the function of the imago, which is to establish a relation between the organism and its reality —or, as they say, between the* Innenwelt *[interior world] and the* Umwelt *[exterior world].*

(Lacan 1949, pp. 1–5)

In other words, the construction of the ego, of *I* and its orientation to a 'reality', is organised around a mirror image of a body embedded in a reflection of 'a little reality', a virtual body situated in a virtual environment. The point here is that, as with Freud's claim, it is the indubitable reality of virtuality that organises not only the subject into a coherent identity but also organises it within a coherent reality.

Lacan claims this image of a Gestalt, an image of unity or wholeness, is at odds with the material being of the infant who is uncoordinated and disjointed, a 'omolette', as Lacan puts it. In assuming the spectral image as its own, the infant anticipates its future psychical and physical coherence but at the same time the image 'splits' the infant between the spectral or virtual and reality. Furthermore, the spectral Gestalt raises the spectre of potential disintergration or fragmentation –of the body in bits and pieces, dismembered.

From this point on, the subject –as a self embedded in a reality –is orientated to or 'split' between two 'competing' worlds: the imaginary or virtual and reality, and two states, being and non–being, that is, coherent or fragmented.

One of the consequences of this assumption of a mirror image of a body as an organising principle of physical and psychical coherence is the persistence of sensation in so–called phantom limbs. And, frequently, images of heaven and hell, in themselves virtual realities, are conjured as states of sublime coherence or alternatively shattering disintegration; and the status or belief in these virtual domains and their position within a cultural discourse is in proportion to the authenticity of representations to the phantasy of coherence and fragmentation.

In each of the above accounts phantasy or virtuality precedes so–called reality and in each case the evidence that substantiates the distinction is an image or spectral reality. From this point of view, reality can be considered a sub–set of the virtual. This would be a trivial point to make were it not that it suggests that the authentic locus of consciousness is within the virtual out of which realities are fabricated and substantiated and that without a prior experience of real virtuality there would be no meaningful reality. It would seem that consciousness is a process, a dynamic that operates in the first instance in real virtuality which subsequently through repression is constrained to operate in a socially sanctioned reality.

However, in electronic environments, the operations of repression tend to be erased and phantasy or the imaginary takes precedence over reality. Consequently, consciousness does not have to constrain itself within the confines of socially acceptable reality and as such generates ever increasing possibilities of being, acting and experiencing that are subsequently translated into reality. In addition, the subject who inhabits the virtual without making a distinction between reality/virtuality is not pejoratively labelled insane because in an electronic environment the virtual environment can be collectively inhabited. This aspect of shared interactive, interrelated phantasy tends to liberate consciousness from the confines of the subjective, interior, private individual –it becomes a form of collective or distributive consciousness. This collective consciousness organised through an electronic or virtual environment impacts on reality in the form of dispersed but collective action as, for example, the subversion of the aims of the World Trade Organisation.

In short, electronic environments have the potential to liberate the processes and dynamics of consciousness from the matrix of a 'reality' –a reality that represses its dependence on a virtual infrastructure. Freedom to inhabit virtual environments means new and different modes of being are constructed which in turn change what is called reality and the subjects who inhabit that domain.

Bibliography

Freud, S., 'Introductory Lectures on Psychoanalysis'. In: J. Strackey (ed. and trans.), *The Standard Edition of the Complete Psychological Works of Sigmund Freud*. London: Hogarth Press, 1953–74.

Lacan, J. (1949). The mirror stage as formative of the function of the I as revealed in psychoanalytic experience. In *Ecrits: A Selection,* trans. A. Sheridan. New York: Norton, 1977.

Notes

1. Throughout the paper I have used the Freudian spelling of 'phantasy'.
2. There are two caveats on the use I make of this material. First, whilst I uphold many of the processes and dynamics of psychoanalysis, I do not necessarily espouse the content these process work on. That is, I think they are valid accounts of modern constructs of gendered subjects –I do not think, in contemporary terms and constructs of the 'post–modern subject', this is now the case.

   Second, I address the material in general terms. It is not possible to give a full explication of some of the processes I appeal to as it is beyond the scope of the paper. In other words, I make selective and simplified use of some concepts and processes analysed by Freud and Lacan.

3.2 Sharing Virtual Reality Environments across the International Grid (iGrid)

Margaret Dolinsky

This paper seeks to answer the following questions: What is the iGrid? What does it mean to share a CAVE Automated Virtual Environment (CAVE)? How are artists shaping experience in a technological world? What implications do sharing CAVEs across a high performance high-speed network have for shifting and changing perceptions?

The iGrid or international grid is a network used by applications that demand high bandwidth for realizing computation intensive states. Some applications include visual displays and some include interaction with persons in remote locations. The iGrid networking effectively allows exploration among many disciplines on multiple levels. From the iGrid 2002 conference website we learn:

iGrid 2002, the 3rd biennial International Grid applications–driven testbed event, challenges scientists and technologists to utilize multi–gigabit experimental optical networks, with special emphasis on e–Science, LambdaGrid and Virtual Laboratory applications. The result is an impressive, coordinated effort by 28 teams representing 16 countries, showcasing how extreme networks, combined with application advancements and middleware innovations, can advance scientific research.

As computational scientists strive to better understand very complex systems –whether biological, environmental, atmospheric, geological or physics, from the micro to the macro level, in both time and space –they will require petascale computing, exabyte storage and terabit networks. A petaflop is one hundred times faster than today's largest parallel computers, which process ten trillion floating–point operations per second (10 teraflops). An exabyte is a billion gigabytes of storage and terabit networks will eventually transmit data at one trillion bits per second –some 20 million times faster than a dialup 56K Internet connection. www.iGrid2002. org/whatis.html

The software, hardware and iGrid infrastructure are used by laboratories and universities that work together to share resources. Many such universities and research institutions are forward thinking into the techno–future and invite artists and collaborators. By combining disciplines, many levels of creativity occur on data networks to create data distribution, information technology, visualization, collaboration and interaction.

At iGrid 2000, the art environment 'Blue Window Pane II' by Margaret Dolinsky,

Indiana University (IU) was showcased among science, engineering and cultural heritage applications. By iGrid 2002, the CAVE applications included the art work:

'Kites Flying In and Out of Space' by Jacqueline Matisse–Monnier, Visiting Artist, the National Center for Supercomputing Applications (NCSA), and the University of Illinois, Urbana, Champaign (UIUC). This piece is a study of the physical properties of flying kites, based on the kinetic artwork of Jacqueline Matisse–Monnier. Such locations as Japan, Canada, France and the USA contributed remote PCs, which control the simulation of a kite. 'Kites Flying In and Out of Space' uses distributed grid computing to allow participants to control the direction of the wind and experience the art and motion of real–time computer graphical kite flying.

CAVE art can also focus on human–to–human interaction across the iGrid. CAVE displays located in Chicago, Illinois; Buffalo, New York; Champaign–Urbana, Illinois; Bloomington, Indiana; and Amsterdam, Netherlands participated in shared environments.

'PAAPAB' by Josephine Anstey, Dave Pape, State University of New York, Buffalo, shares a disco dance environment inhabited by life–size puppets and the navigators' avatars. Navigators can create a unique dancing puppet based on their performance input. They can also go to the dance floor to dance with the puppets and other navigators. 'PAAPAB' focuses on creating interactive drama. Here an immersive story develops through engaging characters in an inviting world design.

'Beat Box' by Margaret Dolinsky, Ed Dambik and Nicolas Bradley (IU) networks audio sequencers and activates graphics for sound collaboration, musical play and learning. Participants use the navigation wand to create music by 'hitting' graphical drums or connecting sounds to intervals on an audio sequencer. The three audio sequencers control percussion notes, bass sounds and ambient loops. The interval represents a distinct moment in a sequence that contributes to the resultant delivery of the collective instruments.

'PAAPAB' and 'Beat Box' begin with a pervasive server in Chicago. The host CAVE site, Amsterdam, launches the complete environment with its avatar. Each remote site enters the environment by activating across the network and connecting its avatar in succession. The server coordinates the information data of every object, avatar, sound and event to update all sites simultaneously across the iGrid. The navigator controls the avatar by moving the location of the body, hand(s) and head. Everyone has a microphone and an echo cancellation system. Participants speak, explore, listen and learn as the projector lights paint a scene for exploration. Each tracked body contributes a character, a personality and a consciousness within a shared environment.

In such groundbreaking 'scientific' events, artists bring visual, spatial and textual languages to the virtual environment. As the arts and sciences share programming languages and exhibition space, high–speed networks allow the arts and sciences to jointly explore new worlds. John Dewey, philosopher, states, 'If all meaning could be adequately expressed by words, the arts of painting and music would not exist. There are values and meaning that can be expressed only by immediately visible and audible qualities, and to ask what they mean in the sense of something that can be put into words is to deny their distinctive existence.' Ultimately the arts and sciences co–exist on the iGrid by combining a foundation of science and art history.

For artists creating collaborative or shared spaces, the iGrid is fertile ground for exploring levels of communication and its play on consciousness. Several CAVEs are now able to communicate dynamically owing to the iGrid's speed and the data's ability to update simultaneous displays. For example, audio is computationally intensive and very difficult to network, often experiencing lag. After several collaborative exhibitions, the iGrid proved far superior –250 times faster to the regular Internet. A high–speed network communication environment of visual interaction and infrastructure technologies melds remote beings in a cyborgian fusion.

Meaning is used in a rhetorical nature from where multiple views and expressions are linked in an episodic fashion. Although the experience in a shared CAVE environment is not linear, its episodic nature is significant in unfolding meanings that are highly contingent upon the exploration. When it begins, the environment is diffuse in meaning. The participants create a narrative as they move through time, describing events to themselves and making choices. If the artist is able to generate fresh and novel associations from participants, there is a potential to shift and change perceptions. Cultures all over the world have investigated technologies in order to develop methods of altering consciousness and even inducing spiritual experiences.

> *Achieving altered states can take place during meditation, hyperventilation, the practice of yoga, hypnosis, fasting, and physical suffering (such as the self–inflicted pain in certain religious traditions or by modern–day body–alchemists). It is a state that can be reached in many ways to explore aspects of reality different from those perceived in an ordinary state of consciousness.*
>
> (Eager)

Strange associations and contextual relationships occur when the artist places participants in a network performance while computers juggle image, movement and speech. In an effort to establish a cyber locale where remote locations present multiple narratives simultaneously, the CAVE houses an environment with several levels of ambiguities. The resultant experience is a stream of people, events and associations, where consciousness changes associatively and dynamically. The CAVE is a strong vehicle for interactive narratives, networked and non–networked. CAVE is a playful, optical device and its networked art situates humans in a virtual environment medium that can unite and ignite multidirectional dialogues between distant places, dreamy geographies and diverse viewers.

The science of optical projections enables the artist to work and conjure the possible relationships in establishing emotion and drama. The artist is building analogies for the CAVE environment as a meaningful creation space. As a result, the artwork allows the user freedom to explore, manoeuvre and experiment between order and disorder.

The iGrid establishes a knowledge maze where academics and researchers have the responsibility to create international dialogues between disciplines. It is not the images, the data or the networks that are new; it is the interchanges between them. As a result, a new literacy is being established that we do not yet know fully how to use.

The iGrid is an international channel for science and culture meetings, uniting people and technologies. Thematically, sources come together from all over the world to communicate more computationally and more visually with one another. While many viewers will see what 'content' that they want to see, the iGrid establishes dynamics for a new literacy, new politics and new associations.

Bibliography

Dewey, J. (1934). *Art as Experience*. New York: Perigee Books, pp. 45–6.

Eager, M. (2002). 'The Shaman Reborn in Cyberspace, or evolving magico–spiritual techniques of consciousness–making, in *Technoetic Arts*, vol. 1, no. 1, pp. 25–46.

Stafford, B.M. and Terpak, F. (2001). *Devices of Wonder*. Los Angeles: The Getty Research Institute.

www.gridforum.org/

www.iGrid2002.org/

www.startap.net/igrid2000

3.3 Interactive, Responsive Environments: A Broader Artistic Context

Garth Paine

Artistic context

As an artist my interest lies in reflecting upon the human condition. There are many aspects of our lives and many facets of the myriad relationships between individuals, cultures, governments and philosophical persuasions that we struggle to find the time to consider in our day–to–day lives. Addressing these issues on some level is integral to the development of understanding and insight into the relevance of differences within the global community.

Art in its many forms –musical, visual, performance and literary –is an appropriate platform for the consideration and expression of these issues. Visual artists, writers, dancers and musicians have explored these issues for centuries with profound results. One only has to look at the prestige attached to the great cultural institutions of the world to see that the product of this artistic endeavour has communal value. It might be argued that the value attached to these institutions is purely financial; that the value placed upon the works they contain is a product of contrary economic principles. Of course, the economics of historical value play a part. However, it would be far too cynical to contribute their communal worth solely to the financial market place.

The 20th century witnessed a movement towards the expression of individual experience. Fuelled by a movement away from patronage towards individual commissions, and artists working from their own inspiration, they offered their works in the commercial market place. This evolution of artistic practice saw the constraints of commissioning bodies lifted and replaced by the concerns of the individual artist.

When discussing the place of the artist after World War II Art, Peter Selz comments:

> *Characterised by an intensely personal and subjective response by artists to their own feelings, the medium, and the working process, it was an art in which painters and sculptors were engaged in the search for their own identity. In the universe described by existentialists as absurd, the artist carried the romantic quest for the self, and for sincerity and emotional authenticity, into a world of total uncertainty.*

(Stiles and Selz 1996 p.10)

He goes on to observe that the 'mechanized mass culture with a plethora of facile and easily accessible public media added to the artist's sense of alienation and the need for individual expression' (ibid p. 11).

These pressures of the changing world led to an art that was driven by self–expression, an art that focused on 'the personal psychology of the artist rather than on the phenomenological world (ibid p. 11).

This development grew out of the main thrust of the avant–garde movements of Europe between the world wars.

> *The surrealists' desire for unpremeditated spontaneity held the promise of true creative freedom ... They worked in a realm of ambiguity and communicated through their gestures an aesthetic of incompleteness. At times this exploration turned towards new and unexpected figuration ... while at others it manifested itself in gestural abstraction. The existential act of making the work was an essential aspect. Even more than previous manifestations of modern art, the dialogue between the maker and the consumer of the work became a necessary element for its completion.*
>
> (ibid p. 12)

The interactive, responsive sound environment seems a logical and appropriate extension of the desires of the surrealist artists. It claims the processors of mass media on the one hand and the sacred space of the art gallery on the other. It combines these apparently opposing forces in an artwork that requires the viewer be the conduit for the creation of a momentary experience. The viewer takes the role, not solely of spectator, but simultaneously of creator; where their gestures create the environment and the environment conditions their behaviour. In so doing they find themselves in a position of contemplation, a position where it is necessary to develop a cognitive map of the relationships between behaviour and environment, action and reaction, individual and communal.

Such a position is, perhaps, the interactive, responsive environment. It reinvents the basis for humanist considerations extending from the physical to the spiritual, the sense of connection to environment, to community, to religion, culture, society and all the other structures that form a broader sense of belonging.

A sense of being immersed in the experience became of paramount importance for contemporary artists. Mark Rothko (1903–70), for instance, spoke of his large canvasses:

> *I paint very large pictures. I realised that historically the function of painting large pictures is painting something very grandiose and pompous. The reason I paint them,*

however –I think it applies to other artists I know –is precisely because I want to be very intimate and human. To paint a small picture is to place yourself outside your experience, to look upon an experience as a stereopticon view with a reducing glass. However you paint the large pictures, you are in it. It isn't something you command.

(Rothko 1951 p. 104)

The sense Rothko had of being immersed in his paintings heralds one of the qualities of an interactive, responsive sound environment. Sound is perhaps the best medium with which to achieve a sense of immersion. It presents as a homogenised sound field but may contain points of spatialised information, points of interest that seem separate, dynamically mobile and yet part of the whole. In order to achieve these objectives, the sound generation algorithms must be designed in such a way that the position of the spectator is considered in the sound spatialisation and that the aesthetic of the sounds reflect an organic and approachable quality.

Robert Motherwell (working in New York at the same time as Rothko) discusses the role of the artist in defining an aesthetic that explores and concentrates emotional experience:

The aesthetic is the sine qua non for art: if a work is not aesthetic, it is not art by definition ... The function of the aesthetic ... becomes that of a medium, a means of getting at the infinite background of feeling in order to condense it into an object of perception. We feel through the senses, and everyone knows that the content of art is feeling; it is the creation of an object for sensing that is the artist's task; and it is the qualities of this object that constitute its felt content. Feelings are just how things feel to us; in the old–fashioned sense of these words, feelings are neither 'objective' nor 'subjective', but both since all 'objects' or 'things' are the result of an interaction between the body–mind and the external world ... It is natural to rearrange or invent in order to bring about states of feeling that we like, just as a new tenant refurbishes a house.

... No wonder the artist is constantly placing and displacing, relating and rupturing relations; his task is to find a complex of qualities whose feeling is just right –veering toward the unknown and chaos yet ordered and related in order to be apprehended.

(Motherwell 1946 pp. 38–9)

While Motherwell is not intending to comment on New Media Art, or more explicitly responsive sound environments, his commentary is particularly pertinent.

He speaks, for instance, 'of an infinite background of feeling' being 'condensed into an object of perception'. It is often my intention to create 'an object of perception' by focussing the visitor's attention on the interrelationship between their behaviour and movement patterns and the quality of environment. The installation work 'is the creation of an object of sensing'. Its entire purpose is to reflect the sensitivity of relationship so that the small intimate gestures of each individual are acknowledged in such a way that every participant is intensely aware of 'an interaction between the body–mind and the external world'. However, as in the case of Rothko's large paintings, it is an 'external world' in which they become completely immersed.

Motherwell's comment that 'it is natural to rearrange or invent in order to bring about states of feeling that we like...' (ibid), describes well the desired nature of engagement with responsive sound environments. The inhabitant is engaged in a constant, fluid and dynamic series of streams of engagement, response and evolution of environmental qualities. This collection of simultaneous experiences, of 'relational structures' establishes an architecture of experience 'whose felt quality satisfies the passions' of those involved in the momentary interaction.

This complex and multifaceted stream of experience must be designed, or at least established as a potential outcome, by the artist. The artist must set out to establish 'a complex of qualities whose feeling is just right –veering toward the unknown and chaos yet ordered and related in order to be apprehended' (Sommerer and Mignonneau 1998 p. 27).

Each of my interactive, responsive sound environment projects sets out to further this aim and illustrate a line of development that represents:

- greater levels of interaction

- more simultaneous streams of response

- dynamic orchestration and

- multi–faceted, conditional response behaviours that reflect more and more intimately the weight of gesture and the quality of behaviour of each individual engaged with the interactive, responsive environment.

This evolution can be seen in my work by examining my interactive, responsive sound environment installations:

MQM (Moments of a Quiet Mind), GITM (Ghost in the Machine), MAP1, MAP2, REEDS AND Gestation, which was exhibited in the 10th New York Digital Salon (April/May 2003) See www.activatedspace.com.au.

Cybernetics – the causal loop

The interdependent collection of 'relational structures' discussed above is the focus of the study of cybernetics, especially with respect to the closed causal loop.

The relationship between the physical space of an exhibition, the technology used to execute the work and the human movement and behaviour patterns that form the basis of individual engagement is critical in the development of responsive environments. These elements form a closed causal loop, that is, a loop in which the only influences are the elements it contains.

Human movement and behaviour patterns act upon the technology. The sensing system analyses the weight of the gesture, the speed and direction of movement and so on. This data is fed to audio and video algorithms that respond in whatever fashion the artist has designed. The response is made manifest in the physical space. It takes the form of changing sound patterns and variations in video or animation projections. In this way the technology acts upon the space, altering the architectural and energetic nature of the exhibition area. These changes cause an alteration of behaviour in those that inhabit the exhibition. This alteration of behaviour, be it one of excitation or placation, will be driven by an intention of bringing the system to equilibrium or to drive it into an unsteady or chaotic state.

The human response to the alterations in the environment forces the closed causal loop into a further iteration. The input to the technology is varied, the output of the technology then varies and the physical space is changed, generating a new and distinct response from those within it.

Interdependent relationships are formed between:

- the technology that mediates the installation response

- the human condition, the behaviour, emotions and relationships that are exhibited in the exhibition space

- the definition and experience of the physical space.

Coming back to my opening comments about communal relationships, I have taken particular inspiration from a key proponent of cybernetics, Norbert Wiener, who conducted extended research into the application of cybernetic principles to the organisation of social systems. In 1948 Wiener wrote:

> *It is certainly true that the social system is an organisation like the individual, that is*

> bound together by a system of communication, and that it has a dynamic in which circular processors of a feedback nature play an important role.
>
> *(Wiener 1948 p. 24)*

These words indicate that a well–designed responsive environment may represent patterns of social interaction and in so doing provide a basis for the consideration of aspects of the human condition.

In 1996 Capra extended this idea when he wrote:

> ... the discovery of feedback as the pattern of life, applicable to organisms and social systems... (helped) ... social scientists observe many examples of circular causality implicit in social phenomena ... the dynamics of these phenomena were made explicit in a coherent underlying pattern.
>
> *(Capra 1996 p. 62)*

So too are the patterns of relationship in an interactive, responsive sound environment made explicit and coherent through many iterations of the closed causal loop discussed above. Each one renders the nature of the relationship in greater detail.

The artists Christa Sommerer and Laurent Mignonneau expressed similar thoughts when discussing interactive digital arts:

> ... the art work ... is no longer a static object or a pre–defined multiple choice interaction but has become a process–like living system.
>
> *(Sommerer and Mignonneau 1998 p. 158)*

One of the pioneers of interactive arts, the American video, and interactive, responsive environment artist Myron Krueger expresses a similar sentiment when discussing his early interactive video works:

> In the environment, the participant is confronted with a completely new kind of experience. He is stripped of his informed expectations and forced to deal with the moment in its own terms. He is actively involved, discovering that his limbs have been given new meaning and that he can express himself in new ways. He does not simply admire the work of the artist; he shares in its creation.
>
> *(Krueger 1976 p. 84)*

Krueger draws the same parallels expressed by Wiener and Capra. He indicates that the experience of engaging in a responsive environment involves an active engagement with each moment, with each moment of engagement contributing to the creation of the artwork. The participant cannot take the stance of a detached spectator; he is inherently part of the process, part of the artwork.

Conclusion

The concept of immersion is integral to the interactive experience. This paper has delved into areas of exploration that parallel and inform interactive research by outlining developments in artistic practice that focus on individual experience but at the same time addressing the cybernetic principles of interdependence.

Much interactive systems research is consumed with the development of technical solutions to particular situations. It is beneficial to take a step back and look at relevant concerns in other fields of endeavour. In so doing I have gained a number of insights that have helped me bend the technology to my cause and not, as is so often the case, the reverse.

Bibliography

Capra, F. (1996). *The Web of Life*. London: Harper Collins.

Krueger, M. (1976). *Computer Controlled Responsive Environments*. Doctoral Thesis. University of Wisconsin, Madison.

Motherwell. R. (1946). 'Beyond the Aesthetic', in *Design,* 47:8 (April).

Rothko, M. (1951). 'I Paint Very Large Pictures: Excerpt from Symposium on How to Combine Architecture, Painting and Sculpture', in *Interiors* 110:10 (May).

Sommerer C. and Mignonneau L. (eds). (1998). *Art as a Living System*. Wien: Springer–Verlag.

Stiles, K. and Selz, P. (1996). *Theories and Documents of Contemporary Art*. California: University of California Press.

Wiener, N. (1948; reprinted 1961). *Cybernetics*. Cambridge, MA: MIT Press.

3.4 Towards Defining the 'Atmosphere' and Spatial Meaning of Virtual Environments

Ioanna Spanou and Dimitris Charitos

Introduction

An issue that has always been of central importance for architecture and environmental design is the way in which structures of feeling are activated through defining and situating the body in space. 'Atmosphere' is defined (Spanou and Peponis 2002) as the objective properties of an environment that metaphorically exemplify structures of feeling through the creation of embodied experience.

Since virtual environment (VE) technology affords the control of environmental parameters for studying human behaviour and understanding, it may also provide a useful platform for conducting experimental studies regarding the concept of 'atmosphere'. It is also understood that the intentional design of atmosphere in a VE could enrich the non–discursive meaning of the spatial experience it affords. In order to identify how this enrichment could be achieved, the issue of atmosphere in VEs may be investigated by studying the environmental knowledge and behaviour of humans experiencing them.[1]

This paper aims to introduce the concept of atmosphere within the context of VEs and at beginning to investigate how atmosphere and the medium of VEs may relate to each other. Firstly, the concept of an atmosphere, the levels of spatial intelligibility relating to the generation of an atmosphere and the manner in which they may relate to causal environmental characteristics and properties are discussed. This discussion takes into account the specificities of space virtual environments and therefore hypothesises about atmosphere induced by both real and synthetic spatial experiences. Finally, reports and observations on the experiences of certain VEs are studied for the purpose of identifying user feedback that describes elements of the experience, which may relate to the creation of a sense of atmosphere from these particular VEs.

The concept of an atmosphere and its relevance to the spatial experience of a VE

In this section an attempt is made to discuss the concept of atmosphere on the basis of relevant literature regarding real environments (REs). In order to develop an understanding of the concept of atmosphere in VEs, however, it is essential to also take into account the intrinsic characteristics of VEs (Charitos 1998).

The concept of atmosphere brings together different levels of spatial intelligibility. At the most generic level, atmosphere bears on orientation, the attribution of importance that underlies our embodied perception of environment. The idea of environmental affordances (Gibson 1986), that is properties of environment that correspond to basic needs of security, protection, comfort, survival, pleasure and so on, has a clear bearing upon this first level. The reading of environment based on such fundamental affordances can be automatic and is likely to influence subsequent symbolically mediated responses.

At another level, embodied experiences acquire metaphorical projections and extensions into abstract meanings in the manner recently proposed by Johnson (1987) and discussed by Lakoff and Johnson (1999). Such metaphorical extensions and projections of embodied experiences are widely shared. They function as fundamental metaphors we think *with*, rather than as metaphors we consciously construct or think *of*. The interplay between the first two levels can account for some of the literal and less literal meanings of the word 'atmosphere'.

At a third level, embodied experiences arise as a consequence of intentional design and as responses to relational patterns that are interpreted as an outcome of design. At this level atmosphere is understood reflexively and implies a conscious awareness of environment as symbolic form. In VEs, where even the most simplistic environments to an extent symbolise real settings or quite often mimic them, we inevitably have a case of a designed environment as a symbolic form.

With reference to real environments, it is suggested (Spanou and Peponis 2002) that whenever a tension arises between the known function of the premises and the visual language deployed in the construction of a building/landscape, this tension generates a particular sense regarding the function and the setting, which is not inherent in the program. If we were to adopt and adapt Frege's (1999) distinction between sense and reference, we might say that the tension between the primary functional reference of the design and the sense produced by the architectural language is obviously fundamental to the meaning that is architecturally ascribed to the place. The sense created by the language of construction contributes to the creation of a definite atmosphere, through the precise way in which it affects the embodied perception of the setting.

With reference to a distinction between signs and symbols drawn by Susanne Langer (1953), it can be suggested that environmental features that are constitutive of feelings work simultaneously as signs and as symbols. Signs situate the body within a field of apparent forces and influences. They correspond to the generic level of a mode of perception, motivated by the needs of an organism, and interpret the environment in terms of positive or negative affordances. The particular features of design that function as signs can also be recognized as symbols. They are part of a design language applied to accommodate a familiar function, which carries its own emotive charges. These feelings are intensely local in the sense that they envelop the subject sensually and not merely symbolically and in the sense that they work at the level of primary affordances and signs.

A question then arises as to how atmosphere is affected by the overall spatial structure of a setting, including the structure of movement, as well as the cognitive retrieval of the structure of configuration.

Symbolic meaning does not seem structured as a narrative. This does not mean that a narrative cannot be constructed. An emphasis upon narrative would imply an emphasis on the logical articulation of symbolic forms and would direct us into looking at architecture as a discursive form (Langer 1979), one where the sequential unfolding of relationships over time is essential to the construction of meaning. The alternative is to think of different moments during a visit at a setting in terms of virtual synchrony, where what is mentally carried from one moment into the next is a memory of atmosphere rather than a memory of precisely articulated form. The impact of patterns of atmosphere, nested into each other, as if in layers of embodied memory, would be to approach each new setting or perceptual horizon, with a perceptual motivation previously felt. Different moments of atmosphere are sewn together as a collage, with partial overlaps, displacements and juxtapositions between parts that retain their independent integrity.

Learning from the impact of a VE spatial experience on VE's participants

Following the discussion above, the actual experience of being in certain VEs is investigated by studying the reports and observations of participants' behaviour within:

- a series of experimental desktop VEs designed for studying navigational and interactional behaviour

- Char Davies 'Osmose' immersive VR installation

This study is expected to reveal aspects of the impact that these VEs had on participants at all levels of the spatial experience and to function as a starting point for understanding how VEs may be designed so as to intentionally induce a certain sense of atmosphere.

Experiments for studying spatial behaviour in desktop VEs

A series of experiments (Charitos 1998) were conducted for investigating the dynamic, navigational behaviour of subjects, their sense of orientation while navigating, as well as their behaviour with respect to relatively static activities located in places, along with the consequent subjective impressions emanating from these experiences.

On the basis of subjects' observations and authors' reports,[2] it is suggested that the type of activity that they aim at accomplishing when entering a VE may influence the manner in which they may experience certain spatial characteristics and qualities of these places. Physical constraints like gravity may be prominent in the manner that participants experience VEs, even when they are consciously aware that they do not apply to the environment they are experiencing. Indeed, in experiments under unrealistic environmental conditions, subjects were classified according to a certain predisposition for navigational behaviour into:

- those who were keen to try out non–realistic conditions of navigating (flying, etc.), enjoyed doing so and did not feel threatened

- those who were still constrained by gravity, felt insecure and were not keen to navigate and act within a non–realistic spatial context

It is not possible to provide a detailed report of the exact environmental parameters considered as causes of these observations. Therefore, this paper will only refer to the recorded impressions/feelings, induced by certain spatial characteristics, in a systematic manner. These feelings may have been induced by a sense of space or by a certain navigational activity and its kinesthetic impact or by a combination of the two.

Metaphorical expressions have been often used for describing these feelings/impressions of:

- a certain space (e.g. 'the secretive place')

- a particular spatial element or characteristic (e.g. 'the cubes felt softer')

At the most generic level of spatial intelligibility, the majority of positive impressions felt by participants related to a sense of security, clear direction and orientation, comfort and freedom to navigate. Negative impressions reported were a sense of disorientation, insecurity, constraint, restriction even to the point of feeling claustrophobic, physical difficulty in manoeuvring for navigation, discomfort, confusion, distraction, disconcertion, uneasiness, frustration, vertigo, shock, threat, fright, fear of heights and a tendency to escape. Other reported impressions which cannot be clearly characterised as positive or negative mainly refer to a perceptual relation of the body to its environment: body awareness, apparent slowness of movement, a sense of rising or falling, a sense of being suspended and a kinesthetic relation of the body to synthetic space within the monitor.

At a higher level of spatial intelligibility, participants reported a series of positive impressions: a sense of comfort relating to a clear definition of space (inside–outside relationship), happiness about navigating in a non–realistic manner, enjoyment, willingness to explore, a sense of comfort relating to a certain sense of place, a positive sense of distraction associated with interest, excitement–curiosity, excitement for experiencing something new and different. Certain negative impressions were also reported: a lack of interest or will to explore, constraint from a very precise definition of surrounding space, an unwillingness for a novel experience, disturbance. Finally, certain impressions at this level could not be classified as positive or negative: a sense of 'unrealness', fascination and simultaneous distraction by a novel experience, pleasant distraction and a sense of being in a 'secretive' space.

It is worth mentioning that even in this desktop VE, subjects often became physically involved while trying to accomplish the task. This involvement was clearly manifested by the movement of their bodies in an effort to manoeuvre their viewpoint by utilising a 3D input device.

Finally, the issue of succeeding types of spaces and consequent experienced atmospheres was discussed. Participants reported certain strong impressions after passing from one spatial entity to a significantly different one. These reports stress the significance of structuring movement within a VE.

Osmose
Char Davies (1998) reports a series of 'unusual' sensations, experienced by participants immersed in 'Osmose':

- a feeling of being somewhere else, in another 'place'

- losing track of time

- a heightened awareness of their own sense of being

- a deep sense of mind/body relaxation

- an inability to speak rationally after the experience

- a simultaneous feeling of freedom from the physical body and an acute awareness of it

- intense emotional feelings, euphoria and an overwhelming sense of loss when the session ends

The assumption that a mobile, wholly–changing environment can be disorientating is put forward by Norberg–Schulz (1971), who agrees with Piaget in that 'a mobile world would tie a man to an 'egocentric' stage, while a stable and structured world frees his intelligence'. Davies's reports on sensations induced by 'Osmose' imply a unique, hyper–real, synthetic spatial experience from an egocentric frame of reference, which may be partly attributed to the dynamic and evolving character of space in this VE. It is, however, certain that these sensations are also largely attributed to the exceptional beauty and ethereal atmosphere induced by 'Osmose'.

Concluding remarks

In an attempt to compare the impact of experimental VEs presented above with that of Osmose, it may be suggested that a VE like Osmose is more likely to induce a certain atmosphere because the motive for its creation is poetic rather than functional, the interactive multisensory experience afforded is richer, more complex, aesthetically more successful and more immersive. Therefore the body participates more in the experience, thus inducing a deeper affective impact and a spatial experience at a symbolic level.

On the other hand, the experimental VEs are very simple, providing minimal settings for the spatial experience to take place. This is due to the fact that they investigate very specific properties of VEs and do not aim at creating any atmosphere as such. Nevertheless, they manage to induce certain strong spatial impressions, mainly at a generic level of spatial intelligibility. It is also significant to mention that participants often used metaphorical expressions to describe their spatial impression.

After reading the observations above, since the experience of VEs fundamentally differs from real spatial experiences and since several participants seemed to adapt relatively quickly to the non–realistic aspects of the experience, it is suggested that there may be a need for developing new sets of affordances that will relate to the generic level of spatial intelligibility in VEs.

Finally, it is suggested that low–level metaphorical extensions are very much at work in both 'Osmose' and experimental VEs.[3] Even though we do not have the ability to extend generic embodied spatial experiences in a VE, owing to the fact that we experience an illusion rather than real space, we may still pass directly to the phase of abstract concepts' projection, metaphorically expressed via the design of the VE. What aids us towards this effort is the awareness of our own body, confirmed by participant reports. Maybe this fact implies a further freeing of imagination for creating spatial meaning.

Bibliography

Charitos, D. (1998). *The architectural aspect of designing space in virtual environments*. PhD thesis submitted to the Dept. of Architecture and Building Science, University of Strathclyde. Glasgow.

Davies, C. (1998). 'Changing Space: Virtual Reality as an Arena of Embodied Being', in Beckman, J. (ed.). *The Virtual Dimension: Architecture, Representation and Crash Culture*. New York: Princeton Architectural Press.

Frege, G. (1999). 'On sense and reference', in Baghramian ,M. (ed.). *Modern Philosophy of Language*. Washington DC: Counterpoint.

Gibson, J. J. (1986). *The Ecological Approach to Visual Perception*. Hillsadle, NJ: Lawrence Erlbaum Associates.

Johnson, M. (1987). *The Body in the Mind*. Chicago: The University of Chicago Press.

Lakoff, G. and Johnson, M. (1999). *Philosophy in the Flesh*. New York: Basic Books.

Langer, S. (1979). *Philosophy in a New Key*. Cambridge, MA: Harvard University Press.

Langer, S. (1953). *Feeling and Form*. New York: Scribners.

Norberg–Schulz, C. (1971). *Existence, Space and Architecture*. New York: Praeger Ltd.

Spanou, I. and Peponis, J. (2003). 'Architectural atmosphere and the spatially situated body', in *Proceedings of the 4th International Space Syntax Symposium*, London (in print).

Notes

1. In this abstract, humans who experience VEs will be referred to as *participants*.
2. All observations and reports taken into account here have either been spontaneously reported by subjects while doing the experiment or while being questioned or have been observed by the author whilst subjects performed the experiment.
3. E.g. in the case of up–down relationships, 'floors–walls–ceilings' definitions, vertical positioning of universes, etc.

3.5 Symbiotic Interactivity in Multisensory Environments

Stahl Stenslie

Currently a silent revolution is going on as an effect of emerging mobile and wireless technologies. In the near future we will not just see technology–driven trends towards the unwired techno–nomad, we will simultaneously experience a fundamental change in the way we merge together with advanced, non–local communication systems. Combined with increased sensory resolution and focus on individual experience, we will symbiotically fuse with a world that, in effect, has become our fleeting interface.

With this scenario at hand I will describe and demonstrate a new form of interactivity, namely symbiotic interactivity. It names the sensory symbiosis of man and machine as made practically possible by the recent developments in wireless communication technologies. It is symbiotic in as far as it renders the historical vision of the cyborg,[1] that is, a functional symbiosis of man and machine, practically possible. The new element now it is the functional implementation on a large scale with standardized technologies. This is one major difference from f.ex. networked experiments of the 1990s.[2] With the implementation of wireless networks like the 3rd generation mobile telephony system (UMTS), we're on the edge of a planned communication revolution. This is made possible by the joint telecom industry's enormous effort to establish the computing mobile phone as the core technology of their business. Interpolating the effects of the recent mobile phone revolution, in Europe in particular, we can expect the future to hold a few surprises in store.

Media influence our perception of reality and consequently change the way we think, act and experience the world. With reference to the impact of the Guthenberg printing revolution, McLuhan pointed to the changes that happen when new media and corporal media–*extensions* are introduced: 'to give the human an eye for an ear is, socially and politically, most likely the most radical explosion that can happen in any social structure' (McLuhan 1962). UMTS is such a new 'corporal' media. An elaboration of McLuhan could claim that these emerging computer–enabled environments will have a greater impact on our culture than all previous media revolutions together. One prediction of this mobile computing revolution is that we will see the establishment of interactive systems built around biospherical metaphors as the new paradigms of interactivity. This is a direct outcome of the perceptual rewiring occurring in multimodal, computer enabled environments.

This rewiring is a result of recent advances in attaching the machine and the human. The PC has not been able to depart from the desktop simply because of its heavy weight alone. That also goes for low-weight laptops. They are even still what they are called: something to sit on your lap and not something you actually use while walking. This limitation in PC usage is mainly caused by the monotone metaphors for digital computing. We still think of and use digital technologies like an office space. We sit in front of the machine and experience the digital world through the outdated metaphor of the camera obscura, contemporarily represented by the heavyweight screen. Rather than being a window to the world, it might be in the way. Since the mid–90s the telecom industry has gone in different directions by developing computing technologies and applications functioning without screens. The success of the mobile phone is based not on its audio–visual capabilities but on its audio–mobile connectivity. Also, mobile phones are personalized technologies. They are small enough to be integrated with your body. This intimate practicality makes your phone number synonymous with your name. When the emerging UMTS networks add hardcore computing technology to their audio capabilities, these personalized, intimate applications will be further integrated. One natural extension will be the inclusion of corporal functions, biometrics and –further along the line –'emotional computing'. Combined with GPS tracking of users' geographical coordinates, we can even expect URLs to be expanded by time and physical space coordinates. My homepage will –for example –only be accessible when you're at Ground Zero.

Reading the body alone is not enough. Actively feeling the world –both the physical and the virtual –will enhance the technological combination of the user with his personal terminal. The scenario of multisensory, audio–corporal connectivity is partly experimental science,[3] partly a future vision played out in artistic contexts (cyberSM 1993; Blast Theory 2002). To investigate these assumptions it is necessary to identify and label dimensions of perceptual experience in multisensory environments. In my work I've employed the qualitative methods of action research to model and predict users' emotional responses. At its core is the active participation of the researcher in the research process where changes are actively induced.

My 'Erotogod' experiment[4] is one example of such research. It is an experimental, immersive and multisensory media–art installation that probes into the boarders of emerging technologies.[5] It elaborates on the sensual dimension of aesthetics by connecting advanced computer and interactive technologies to the body as a sensual interface. In 2003 it was presented at the Dutch Electronic Art Festival (DEAF). This practice–based art installation is a multisensory space of experience that lets the user interactively write his own myths of creation by touching his own body. Upon entering the installation the user first kneels down in order to have his head placed directly inside an audio–sphere consisting of 16 loudspeakers. The spherical positioning of the loudspeakers pointing toward the user's head creates a physical 3D and 16 channels audio system. The real–time sound composition is written with Max/MSP running on a Macintosh G4 1.2 MHz with 16 channel audio output.

The user is then dressed in a bodysuit hardwired to the installations special interface.[6] In effect, the suit turns the body into a sensory instrument. By touching his own body the user can influence the overall state and expression. The bodysuit has a two–way functionality, including both sensors and effectors. Placing sensors directly on the body presents problems, such as involuntarily triggering the system through movement and general squeezing; in addition there is the time–consuming process of placing the sensors in the same position for each user. Mechanical sensors were custom built to be relatively insensitive to accidental movement and, at the same time, require the user to push to decisively to signal 'I am touching'. The system, therefore, recognized fairly well a touch signal that stemmed from an intentional action. Altogether 96 sensors were placed in a grid system inside the suit to cover the body. Underneath each sensor we placed one custom built, vibrotactile effector. These each had a variable output ranging from 'soft' to 'intense'. They were made strong enough to provide the user with the sensation of 'touch' even when he was wearing his own clothes underneath. Such 'spots' make it easier to map the body and secondly they allow for direct feedback. If the user presses a button the vibrator underneath gives him immediate feedback. This made the autoerotic functionality of the suit more intuitive.

Additional effectors were placed on around the body, allowing the installation to 'draw' continuous patterns of touch and to haptically immerse the user. Now, comparing the sensory resolution of a hundred sensors and vibrators with the millions of various receptors of the body might not seem fair. However, practically speaking, the 110 effectors used covered the body fairly well. Each effector measured approximately one by two centimeters but in relation to the senses, it covered an area of at least five by five centimeters. With such 'real' strength of each effector I could have implemented up to 200 but for the basic range of sensory sensation it would not matter much qualitatively. It is much more critical if the number of effectors is decreased. Below 30 it is hard to induce impressions of full body sensory immersion.[7]

The challenge posed by the bodysuit was how to make sense out of a hundred sensors and effectors. One hundred individual buttons can generate an enormous amount of data. The system design was simplified by grouping all sensory spots into five 'intelligent' zones. These were one for each leg, one for the groin, including the back, and two for the chest. This grouping made sense both to simplify the construction of the system as well as to make using it more intuitive. Throughout the project we have continued the development of a haptic language. There is no intersubjective, general code for understanding what a) pressing my upper right chest should mean and b) what a vibrotactile (pulsating sensation) in the same region implies. Such problems were continuously encountered and worked out. From the Human–Computer–Interface (HCI) point of view, one of the success criteria was for the users to dress in the bodysuit, engage with the system and create an intuitively interesting experience within the timeframe of ten minutes. This time span was short but necessary for the installation overall to function within the context of an art exhibition. It was also sufficient to observe the user's initial responses to the multisensory experience. From observation of the users that returned for a second or third time, an increase in the qualitative impression of system immersion was observed . Ideally the time span for each user could therefore have been doubled but it was not critical in this setting.

By touching his body the user uploads a 'sensitive' imprint of his body into the machine. In addition to influencing the sound and corporal sensation, the installation used these raw, sensory data to output a 3D real–time, open–GL based rendering of a graphical, text–based universe. This was seen on the three screens surrounding the user. The installation drew graphics based on actual words and letters. The front projection focused on creating a dynamic and graphical textual expression that did not necessarily have to be read as text. The actual text the user produced was projected as readable words and sentences in a second 3D rendering on the two side walls. All expressions of the system took shape as a cumulative result of the system–user interaction. The content of the texts were re–combinations of excerpts of texts taken from the religious creational myths of the Torah, the Old Testament and the Koran. The recombined text was a result of 1) user activity, 2) the state of the system, 3) timeframe position and 4) system 'intelligence'. The new, user–based myths appear as three–dimensional sound, graphics and corporal experiences. Together these expressions make a synesthetical experience, that is, the combinations of multiple sensual modalities creates a larger impression.

The degree of usability is critical. The use of the bodysuit was relatively transparent, even with a short time span per user. Developing a multisensory language is harder. What does the input and output mean? What does touching oneself 'there' mean? And what effects does it produce? The best results were obtained when a user first encountered the closed bodysuit/environment through a one–to–one reflection: when he f.ex. pressed the upper right breast, he heard sounds from the upper right part of the audio sphere. This one–to–one, body–to–system analogy functioned well as an introduction to using the system. To measure the user experiences I employed the concept of perceptual breakdown.

> What I can name cannot really prick me. The incapacity to name', however, 'is a good symptom of disturbance.

(Barthes *in Camera Lucida*)

As Barthes notes, in an aesthetical context disturbance is a highly charged experience. The perceptual conflict between sensing something and not knowing it is –in a platonic sense –an erotic experience. It propagates a desire to achieve that which one senses but does not yet possess. Barthes also suggests that this is a creative dimension. It represents a challenge to our habitual `being in the world' (Heidegger). It is also a distinct sensation. You know when you're irritated. In this sense it also represents a qualitative measurement. An interesting perceptual test system (art work) should be able to induce user–sensations that alternate between the breakdown of (user) expectancy and satisfaction. Breakdown situations are described both in activity theory and phenomenology. Heidegger introduces the concept of breakdown in his book *Being and Time*. When I have the hammer in my hand and hammer a nail, the hammer does not appear to me as a separate object. It is only when a breakdown occurs –as when the hammer falls apart –that the hammer appears to me as something separate from me. In the experiment with symbioactive interactivity such breakdown situations were introduced by saturating the user with multimodal, sensory–stimulus.

By active user observation, video analysis, questionnaires and interviews I found the users to actively engage in the self–reflective communicative process represented by the installation's sensory loop. Their experience was one of surprise, physical enjoyment and engagement. Direct, perceptual breakdowns as in the dysfunctional hammer scenario were not clearly observed. However, for many users the immersion in sensory experience represented a breakdown of expectation. This can be viewed as an indication of symbiotic interactivity.

Within the framework of Consciousness Reframed, I find Heidegger's concept of 'intentionality' useful. It describes our being–in–the–world as a continuous process. There are no final answers, no right or wrong, but a continuous engagement with the world. This represents a way to deal with such abductions caused by symbiotic reduction. Even if near–future interaction technologies reduce the importance of the individual, they can be also be used to fuel intentional, dynamic processes –and so might the fundamental modus operandi keep us conscious. This consciousness can also be expressed as indirect, tacit and body–based: the notion of phatic communication as a crucial constituent in the coming, wireless communication society. The phatic connotes the use of so–called 'empty phrases' like 'how do you do, ' 'how's the weather', etc. Even if seemingly purposeless, the phatic dimension serves a very special purpose, namely to keep us in contact. This sense of connectivity is essential to unleashing the full potential of communication. The phatic dimension emerges in multisensory, symbiotic interactivity facilitates for our being–in–the–world as a continuous process.

Bibliography
McLuhan, Marshall (1962). *The Gutenberg Galaxy*. Toronto: University of Toronto Press.

Notes

1. Haraway, *Cyborg Manifesto*; Gibson's 'Neuromancer', etc.
2. Knowbotic Research, Stelarc, my cyberSM experiment, etc.
3. remote telehaptic research as in controlling space stations
4. Homepage www.stenslie.net. Undertaken with Knut Mork, Trond Lossius and Asbjørn Flø.
5. 'emerging technology' as of 2003
6. Custom–built interface with 128 analogue outputs, 96 digital and 16 analogue inputs, run by a Lua server
7. Based on my experiments with the 'sense:less' system, see www.stenslie.net/stahl

3.6 Aesthetics within Ego Shooter Games

Maia Engeli

The search for media–specific aesthetic expression in virtual worlds is essential. Currently these worlds are dominated by simulations, replications and metaphors from the known physical world or by fantasies of a physical reality. The language of virtual worlds has not yet developed very far and the 'essential properties that will determine the distinctive power and form of a mature electronic narrative art' (Murray 1997) have not yet been fully identified.

'The success of the fighting contest games poses a challenge to the next generation of digital artists' (Murray 1997). Ego shooter games are the most popular applications within the genre of virtual worlds and the most widespread of them (Quake, Unreal Tournament) are largely open–source and offer editors for level design. Their openness and popularity makes them a media of choice to explore aesthetic possibilities. Their popularity is furthermore reflected in the existence of a large, multifaceted community. Members play games together, discuss different issues, exchange knowledge and data, engage in gossiping and make creative use of the possibility to create their own players and virtual worlds with the provided editors. There is an audience and it is not only artists who are aware of it: the commercial world is detecting its potential for marketing. For example, via product placement, or the creation of his own games on top of an existing game engine, like 'America's Army', a free first person shooter game invites you to 'Become a member of the world's premier land force'. (America's Army 2003).

'Because guns and weapon–like interfaces offer such easy immersion and such a direct sense of agency and because violent aggression is so strong a part of human nature, shoot–'em–ups are here to stay. But that does not mean that simplistic violence is the limit of the form' (Murray 1997). Jenkins and Squire observe that gaming is not about 'realism but rather 'immersiveness'' and 'As a consequence, a close consideration of game space reveals a broad range of aesthetic influences, including Expressionism (which maps emotions onto physical space) and Romanticism (which endows landscapes with moral qualities). Surrealism is another modern art movement that has influenced game design. The Surrealists created dreamlike images which nevertheless followed the conventions of representational art, often deploying familiar stories (such as those in the Bible) as a basis for psychologically complex, symbol–laden environments' (Jenkins and Squire 2002).

In games the expressiveness is composed from visual, aural and interactive aspects in a dynamic digital space. Aesthetics beyond the visual have to be considered, including structure, experience and meaning. Meaning is primarily evoked by perception through interaction, while reading words or iconographic images is sometimes used as a secondary means to disambiguate the intended message. There is also no clear culturally based language with a long tradition, therefore the messages are more ambiguous. But they can be understood thanks to the increasing 'maturity' of the player, who has progressed from a reader of texts or a viewer of images to a user and now a player. Or, as De Kerckhove describes it, 'This is the passage from what I define as 'point of view' to 'point of being': the 'point of view' expects for each body to observe the world from its interior; the 'point of being' implies an intense exchange between a single individual vision and the multiplicity of exterior visions' (Kerckhove 2003).

The process of generating levels within ego shooter games is a design and implementation process with media–specific characteristics as well as different emphases depending on the background of the creator. Important differences to other design processes include: 1) Levels are built on a 1:1 scale; they are real and not simulations or mock–ups. They are perceived from within at every stage of the design process; the god–like observer perspective –as known from other design processes –may only be an option if it is deliberately chosen. 2) Levels can be modified. The design and implementation process is characterized by feedback loops, which naturally can include feedback from various players joining and testing the level over the Internet. 3) Various dynamic and reactive aspects are available and can be visually programmed. 4) There is a mix of natural and artificial intelligences that can inhabit a level. The introduction of the artificial ones is partially in the hands of the creator of a level. 5) It is important to remember that a shooting game is a shooting game. It is very hard to modify it into something totally different. It is much more efficient to keep the shooting as the shooting game's primary driving aspect and to integrate the shooting aspect within the goal of the level one is building.

Ego shooter levels and modifications by artists

'NextLevel' (by Miriam Zehnder, Eric van der Mark, Patrick Sibenaler, ETH Zurich, http://caad.arch.ethz.ch/~patrick/LOCAL/research/playground/) investigates the relationship between game, architecture and narratives. It is a reinterpretation of the novel *Through the Looking Glass* by Lewis Carroll. Alice, the story's main character, imagines different worlds for the different characters of the chess game. The design of the game level was not done as a direct translation of the novel but aimed at 'understanding the essence of the idea and then creating meaningful interpretations' (Zehnder 2001). The environment is hybrid and hyper–linked; it contains contrasts on the visual as well the behavioural level and includes a broad variety of visual aesthetics, from the common game aesthetics to very abstract, reductionist representations. This makes the journey through Alice's world a real adventure with unpredictable encounters not only on the visual level: gravity does not always function as expected; speed of movement is relative; or what seems to be an enemy may turn out to be one's own mirror image.

Jodi's 'untitled game' (www.untitled-game.org/), game levels in which the interface has been reduced beyond the minimum, is a 'subversion of the aesthetic expectation of opaque source codes and understandable output' (Cramer 2002). The reduction of the visual aspect to black and white 'pixels' provokes a distorted perception of the movement in the game. Nonetheless, the dynamics remain the only level that can be interpreted and after some time one can read the motion patterns as familiar from shooter games. The effect is a heightened awareness of these shooting patterns as well as the insight that even this most abstract representation can be as terrifying as the realistically rendered game.

Max Moswitzer and Margarete Jahrmann dig deep into the possibilities to alter the game and combine it with other information sources. In their newest example 'Nybble Engine' (www.climax.at/) they cracked the server part of 'Unreal Tournament' to replace every action with a procedure and therefore use the engine as 'material to which engineering can be applied' and special messages created. 'An aesthetic message is usually the deconstruction of a conventionalized text form or a media text. It is receded by destroying semantic portions in order to increase the aesthetic information' (Jahrmann and Moswitzer 2003). For example, shooting has been altered to send anti-war emails to the president of the United States.

Ego shooter levels from workshops with architecture and new media students

'Dreamday' (by Marc Dietrich and Michael Huber, June 2002, New Media, Hyperwerk, Basel, Switzerland) is built as an absurd, kitschy, two-story house filled with cliché loaded images providing a dream world that can be leisurely explored. The atmosphere is just too pleasant so that playing a death match within this too beautiful environment is a welcome contrast.

'Darwin is Dead' (Anja Kaufmann and Julia Kehl, June 2002, New Media, Hyperwerk, Basel, Switzerland) is about emphasizing chaos theory over Darwin's theory. Through the specific use of textures and the use of mirrors, the shape of the space dissolves into a visual pattern enhancing its labyrinthine characteristic. At some specially marked locations the player is teleported into a different space, patterned with fractal images. The more of these spaces that are visited, the more of the Darwins disappear on the centre volume and are replaced by patterns representing chaos theory.

'Hidden Islands' (March 2003, Architecture, TU Delft, Netherlands): In this level the city does not show its parks but the birds give you hints about where the green islands are hidden. Each of the five parks represents a typical 'park' theme, such as 'meadow with trees', 'zoo with rabbits', 'a sports team', 'a playground' and 'a pond'. All living beings in this level can be shot, the consequence being the spreading of nervousness and disorder as well as items being out of place.

'Golden Calf: Metareal' (by Luca Vincente, June 2003, New Media, Hyperwerk, Basel, Switzerland) is an immersive interactive exhibition of thesis work. In addition to the expandable exhibition space, there are spaces that illustrate the often painful process of the thesis work. There is the 'nimbus' where the player floats comfortably but without control of direction and slowly gets desperate to find an exit. There is 'hell': the player falls down and sees nothing, is lost in the dark until he recognizes a dim light far away. And there is 'heaven' with windows down into the exhibition. The windows show different perspectives onto the works. Visually the level is abstract and mostly black and white deliberately displaying a contemporary graphical aesthetic.

Conclusions

In their search for a media–specific aesthetic for virtual worlds, creators are fighting with the limitations of the editors, the programming and the available graphics libraries but also with the acceptance by the players and their abilities. Since the player is such an important factor, the popular ego shooter games are an acceptable means to explore aesthetics for virtual worlds. Nonetheless there is a trade-off between the fact that they are shooting games and using them because of their popularity. Other virtual worlds that expand the chatroom metaphor or focus on information display certainly allow for a different freedom regarding their aesthetics.

Bibliography

America's Army www.americasarmy.com (visited 17 July 2003).

Cramer, F. (2002). 'jodi.net', in install.exe/Jodi, Baumgärtel, T. and plugin (eds.), Basel: Christoph Merian Verlag.

Jahrmann, M. and Moswitzer, M. (2003). 'Nybble–Engine', Vienna –more?

Jenkins, H. and Squire, K. (2002). 'The Art of Contested Spaces', in L. King (ed.). *Game On*. New York: Universe, pp. 64–75.

de Kerckhove, D. (2003). 'Searching for the Principles of Web Architecture', in F. Barzon (ed.). *The Charter of Zurich*. Basel, Boston, Berlin: Birkhauser –Publishers for Architecture, pp. 38–67.

Murray, J. H. (1997). *Hamlet on the Holodeck –The Future of Narrative in Cyberspace*. New York: The Free Press.

Poole, S. (2000). Trigger Happy. Videogames and the Entertainment Revolution, New York.

Zehnder M. (2001). 'nextLevel', in M. Engeli (ed.). *Bits and Spaces –Architecture and Computing for Physical, Virtual, Hybrid Realms –33 Projects by Architecture and CAAD, ETH Zurich*. Basel, Boston, Berlin: Birkhauser, pp. 198–201.

3.7 Creative Communities in Networked Hybrid Spaces

Mauro Cavalletti

Creative communities in networked hybrid spaces

Current re-evaluations of the role played by traditional physical elements in information technology systems have challenged one of the most significant paradigms of the information age: the idea of a society where materiality would be diminished in its intrinsic value. The once mostly digital domain of systems –the many de-materialized relationships made popular by the Internet –is being transformed by heterogeneous interlaced networks which make use of personal objects, cloths and elements of living and working spaces in our surroundings as network nodes, connections and interfaces. The relatively controlled environment of interconnected computers is quickly extending beyond itself to objects of our daily life, generating a new breed of webs we are only starting to discover.

This fresh hybridization of new and traditional systems generates opportunities for a whole set of investigations and offers a considerable potential to be explored in creative approaches. In order to design information networks that can operate in this complex and more intricate context, we now have to reframe our relationship with these highly informative environments. We need a better understanding of the information relationships the transformed objects and elements in our surroundings contain, relearning how to interact and how we might communicate with other people through their use. Investigating the design of hybrid spaces as interfaces of larger networks suggests a promising field among the multiple significant opportunities opened by this convergence.

This paper discusses how the mix of information technologies and material elements in architecture can support conditions to facilitate the emergence of creative insights and the distribution of knowledge in the context of large-scale information systems. It will introduce the Creative Network, a scenario-based concept design describing an interactive environment where intelligent networked spaces support the dynamic flow of ideas as well as the physical flow of nomad professionals, while enhancing creative processes and facilitating the development of new knowledge. It works as a high-level formulation, a broad proposition aiming to define a large scope of investigations where future pieces of design and more detailed conceptual work can be further aggregated.

Some points of investigation proposed in the context of the Creative Network presented are: the architecture of the backend system for dynamic sharing of information; how living/working spaces can be prepared to enhance this dynamic and at the same time be adaptive to the changes implicit in this flow; how these spaces can learn with humans, storing and communicating knowledge with their occupants and the rest of the network; and how these spaces can diminish transition impacts and provide optimal experiences for nomad professionals.

The Creative Network

The Creative Network consists of an international community of creative professionals who make use of intelligent working/living spaces spread across the world interconnected by dynamic database systems for knowledge organization and sharing. This fictional network is a creative environment supporting increasing levels of collaboration and originating streams of emerging ideas in decentralized organizations. In its context, creative investigators such as scientists, artists, researchers, designers and architects are network dots creating ideas and concepts in different slots of time and space. Likewise, the ideas and discoveries generated by this community move dynamically in this network. Its members are dispersed, a new kind of nomad, aggregating in diverse places over large geographic areas. Physical environments specially prepared with hybrid components function as clusters and accelerators for creative production that is dynamically shared throughout the entire network.

Contrary to the defined focus and limited scope of knowledge management investigations sponsored by large corporations, the Creative Network can offer an environment for more complex formulations. First, it is based on broad and diverse groups of users, whose common point is the exercise of creative processes, practices that are divergent by nature. Also, creative professionals do not generally make a clear distinction between their professional and personal lives, rendering the boundaries between working and living spaces fuzzy. Finally, creative professionals are always trying to balance new challenges and discoveries with a level of comfort necessary to maximize the creative flow. This makes any attempt to encapsulate or control processes a difficult task.

Owing to the nature of the information produced by the members, the relationship with their surroundings and the ever–changing processes they apply, the environment of the Creative Network should reconfigure hierarchical organization and would increase the possibility of raising concepts, unpredictable work processes and cutting–edge ideas. The system supporting the Creative Network can enhance new tendencies in the form of interest clusters shaped independently from centralized decisions. This formulates a new economy where ideas, trends, insights and discoveries are currencies. This demands more sophisticated yet simpler solutions in order to organize flows in the context of the network.

The construction of the Creative Network, however, does not ask for a complete organization built from scratch and installed at once. It instead makes use of existing assets and proposes an enhancement of actual information sources. It augments the traditional functionality of physical spaces by the improvement of current interfaces and use of new protocols and network technologies.

Setting up the backend for the Creative Network

The first step in setting up the Creative Network is to prepare its backend –the system of servers, dynamic databases and knowledge–share software that sustains the stream of ideas and organizational information. This comes first because it builds on existing networks supported by institutions or interest groups, or temporary project structures, that intend to have their findings communicated and their information sources expanded. A system flexible and intelligent enough to support the storage and traffic of a diverse set of document formats reorganized according to dynamic parameters can leverage existent databases. Further, it can provide critical mass to precipitate a significant exchange of ideas in the network.

The Semantic Web proposed by Tim Berners–Lee is optimal for enabling data sources to be reorganized dynamically depending on the interest of users or systems. Generally speaking, documents can be grouped in the backend by semantic associations, generating ontologies and making the information understandable by machines. This means that, according to preset parameters, large amounts of data can be organized with reduced direct users' manipulation, once complex and demanding associations can be operated by software. In the case of the Creative Network, it means that a number of hybrid spaces and intelligent physical elements in architecture can also be designed to organize and communicate information, helping members to perform their sophisticated associations.

Reducing the Information Architecture provided by a semantic web system to the simplest application in the context of the Creative Network, any kind of digital document or creative unit could be stored and associated with a new layer of attributes generated to qualify it. This layer that can be updated without changing the original document, providing the information used to reorganize the sources accordingly. Any time a new parameter is provided, the documents are reorganized in a new context.

Applying these characteristics to a time/space sensitive event, such as a conference, for example:

As soon as a participant feeds the system with a paper she could immediately gain knowledge, through a simple set of rules such as the association of key words, of other submitted papers related to her interests. She could, for example, map the connections, read them before the conference, contact authors or verify their developing production. In some ways this prepares her for making the best use of actual contact, diminishing time and space constraints. Using the same source in a different organization, the conference promoters could better understand how participants form interest clusters, how they interrelate and even track the evolution of the group.

Expand the same metaphor to a larger number of events, institutions and places; open it to a broader range of interests with considerable diverse participation and we can picture the potential richness of information flows in this context.

Enhancing dynamics by the use of hybrid spaces

Transporting the same line of thought from a mostly virtual dimension to a more physical layer, nomadic individuals carrying knowledge from part to part at a very high pace considerably increase the possibilities of cross–pollination amongst different groups. If we consider the scale of individuals and groups moving throughout interlaced paths but still permitting that clusters of similarity are organized and revealed, we have a really interesting fabric of possibilities where before we would have mere frictions and collisions.

Using intelligent spaces with large–scale collaborative interfaces as nodes in this ever–changing network will intensify these possibilities in a tremendous way. Applying very simple information systems to augment the functionality of spaces, enhancing creative activities and recovering knowledge would be an interesting and challenging starting point for architecture investigations. Applying these investigations to a context of virtually interconnected places separated physically by long distances is even more interesting. The combination of augmented local experiences, increased information flows and mobility in a new type of territoriality generates a very particular situation.

In the current dynamics of the networked economy, places for creativity have functioned more and more as a temporary settlement for nomadic individuals in search of the best support, the most interesting subjects for their creative investigations or simply the best creative minds to team up with. Through the information sources in the network, individuals can be informed about activities or achievements of other clusters. They can track if a certain location (or institution) is or will be hosting a group working on their own interests at some point, planning relocations and subject exchanges. As nodes in a system that facilitates information exchange, the working/living spaces in the Creative Network provide enhanced physical support for this movement. Its structure, taking advantage of a diversity of inputs and divergent approaches, supports and moreover encourages mobility –in this environment all parts benefit from the increased interchange of ideas resultant from high turnover levels. This scenario generates a new type of territoriality that occurs in a sort of trans–urban space, physically unlimited. These locations form a virtual neighbourhood spread across the globe.

Spaces can learn, store and communicate knowledge

Historically, physical proximity has played an important role in supporting focused investigations, collaborative work and discovery sharing. Centres of investigation, residences for artists, research laboratories, design studios and art schools have worked as physical aggregators of creativity. They have always functioned as destinations where people bringing their own skills and ideas converge, finding support and resources to shape their thoughts. In that sense, every place has its own identity and leverages creative outputs with a mix of top–down directives and accumulated learning from bottom–up experiments.

As they can retrieve and organize ongoing discoveries and maximize feedback, the working/living spaces in the Creative Network empower creative processes, enabling individuals and groups to maximize performances and quickly communicate results but also to dynamically redefine and evaluate completion and to reformulate goals. The environment supported by the Creative Network moves the creative process focus beyond a predefined target to broader objectives, pushing potential results to a new level.

In a certain way, these intelligent working/living spaces would act as catalysts themselves, as living elements, each one with its own characteristics and personality as its past (and future) uses reshape the creative dynamics they support. In fact, as aggregators, creative spaces accumulate considerable levels of creative energy and frequently support the intense exchange of ideas. In the most successful situations, this exchange is so high and the levels of energy so powerful that creative opportunities are increased. Very often, new skills are developed and promising ideas are transformed into unpredictable achievements. However, most of this creative energy moves towards entropy after dispersion of groups or at the end of a formal period of time. Preparing spaces to retain information, store it, reorganize and retrieve it, according to new uses, is a way to allow for continuity and proliferation.

Using the dynamic backend in the working/living spaces to store and reorganize information according to context, and hybrid architectural elements as intelligent agents, any kind of creative production feeding the local system can potentially start a dialogue with discoveries and ideas produced any time in that locale. It makes communication with past and future occupants possible simply by the use of the space.

Diminishing transition issues and supporting optimal creative flows

One of many interesting possibilities in this scenario is the enhancement of creative flows experienced by members of this community. A commonly reported feeling by creative nomads is the difficult adaptation into a new space, a sort of transition period when the excitement of a new environment cannot yet be channelled into new output because of the lack of specific elements necessary to the continuity of their creative process. Such elements might be tools, references or simply certain ambiance or light conditions.

The networked working/living spaces can be prepared for optimal experiences, on the one hand by diminishing psychological or physical barriers and on the other by opening digital channels where important elements for these practices can find a free traffic. Long–term architecture programmes could be imagined as continuous temporary arrangements, preparing flexible layouts using reconfigurable elements. This would give members the ability to have some level of control over the space according to their needs and preferences. In turn, this reduces transitional impact of relocation. At the same time, the possibility of moving a considerable part of a member's belongings, tools, sources of reference and inspiration to the virtual space of digital storage would permit anytime access to knowledge at any point in the network, populating the spaces with personal and familiar elements. That combination could provide the Creative Network members the possibilities of plug–and–play processes of optimal creative experience as well as a sense of continuity when moving from place to place.

As the use of these sources can be proportioned in an amplified way by the hybrid architectural elements (such as interactive displays, information walls, sound and image systems or any sort of interface integrated in the physical space), breaks in the flow are drastically reduced. At the same time, the environment is totally transformed by the presence of a newcomer or an ongoing project, as the information retrieved is always an active element in the space.

Open source environment for collaborative investigations

The experience proposed by the Creative Network provides an optimal environment for several levels of collaborative investigations on both design and use of intelligent spaces and networked communications. From the macro organization of knowledge to the most granular development of hybrid interfaces, members of this community are at the same time extreme users and creative developers. This set–up, by definition, can be transformed in a large–scale open source investigation, a productive environment where complexity and simplicity are blended for innovative development of technology.

Bibliography

Berners–Lee, T., Hendler. J. and Lassila, O. (2001). 'The Semantic Web', in *Scientific American*, May 2001.

Cavalletti, M. (2002a). 'Creative Network', in Serve City, Bauhaus Kolleg III, 2nd trimester. Dessau: Bauhaus Dessau Foundation.

Cavalletti, M. (2002b). 'Communicating Knowledge in a Network of Projects', in Serve City, Bauhaus Kolleg III, 1st trimester. Dessau: Bauhaus Dessau Foundation.

Cziikszentmihalyi, M. (1996). *Creativity: Flow and the Psychology of Discovery and Invention*. New York: Harper Collins.

De Landa, M. (1997). *A Thousand Years of Nonlinear History*. New York: Swerve.

Grzinich, J. (2002). 'Mediated Home Space', in Serve City, Bauhaus Kolleg III, 2nd trimester. Dessau: Bauhaus Dessau Foundation.

Ishida, T. and Isbister, K. (eds.). (2000). *Digital Cities: Technologies, Experiences, and Future Perspectives*. Tokyo: Springer.

Johnson, S. (2002). *Emergence: The Connected Lives of Ants, Brains, Cities, and Software*. New York: Touchstone.

Lemaire, P. (2002). 'Project Spaces', in Serve City, Bauhaus Kolleg III, 1st trimester. Dessau: Bauhaus Dessau Foundation.

Mitchell, W. J. (1996). *City of Bits: Space, Place and the Infobahn*. Cambridge: MIT Press.

Montilla, A. (2002). 'The Abhittipura Project: Service Provided = No Border'. Paper presented at the Conference 'Consciousness Reframed IV, 2002: Non–Linear, non–local, non–ordinary', at the Curtin University of Technology, Perth Australia.

3.8 From Multiuser Environments as Space to Space as a Multiuser Environment: Cell Phones in Art and Public Spaces

Adriana de Souza e Silva

Our concept of the Internet is changing owing to the emergence of mobile technology devices, such as cell phones, PDAs and laptops. During the last decade, connection to the Internet was achieved mostly by means of a desktop computer and cables connected to the telephonic network. These interfaces detached virtual space from physical space. Today we identify a hybridization of space, since virtual is coming closer to physical.

Mobile phones virtualize space by enfolding distant contexts into the present context. The emergence of mobile and smaller interfaces to connect to the Internet are responsible for a hybridization of space, blurring borders between physical and digital. In addition, the act of moving through physical space is similar to inhabiting a virtual environment, since it is possible to connect to people who are not present but, even so, change the nearby context. Traditionally multiuser environments have been considered sociability places, which in order to be inhabited needed to be detached from physical space. On the other hand, urban space becomes a multiuser environment whenever mobile interfaces bring the virtual into public spaces. The isolation of cyberspace in the last decade required users to create avatars in order to represent the body across the screen. Since one could not really be on the other side of the screen, it was necessary to redesign the physical body. Nowadays, nomadic technologies bring virtual closer to physical and users turn into walking avatars, moving constantly between both instances.

Multiuser environments as (virtual) spaces

Multiuser environments became popular for allowing many people to connect to the same virtual place at the same time. Certainly multiple people are able to access the same website simultaneously but a multiuser environment is defined when these users have the awareness of each other's presence. Therefore issues about presence, activity and identity have been critical to studies on these places during the last decade (see Turkle 1995, Murray 1999, Rheingold 2000, Donath 1998).

MUDs (MultiUser Domains) are virtual places with textual, 2D or 3D graphic interfaces in which users (inhabitants) should choose avatars in order to interact with each other. Users are sometimes also able to build these environments, which are usually (literal) representations of the physical world. Owing to the fact that

MUDs are virtual environments, they have been considered immaterial places. This aspect is a consequence of considering cyberspace an information space. According to N. Katherine Hayles, in 1948 Claude Shannon defined information as 'a probability function with no dimensions, no materiality, and no necessary connection with meaning' (Hayles 1999 p. 18). Consequently, the concept of information was completely disconnected from the material interfaces in which it is (necessarily) inscribed. This idea contributed to the move to adopt the word *cyberspace* to describe the *Internet*. When William Gibson coined the term cyberspace (Gibson 2000) he was influenced by the concept of cybernetics, which primarily considered humans as 'information–processing entities who are *essentially* similar to intelligent machines' (Hayles 1999 p. 7).

With the intention of discussing the ongoing concept of the virtual and its relationship to materiality, Itaú Cultural, a cultural institution in São Paulo, Brazil, developed in 1999 a media arts exhibition called *Imateriais* –not surprisingly, 'immaterialities'.[1] A 3D multiuser environment with the same name, created by Jesus de Paula Assis, is an example of how MUDs were considered (virtual) spaces and how its author challenged some basic issues related to these of environments.

Imateriais aimed to take physical sensations to the virtual space. One of the creator's goals was to show visitors that the virtual was progressively being assimilated into everyday life. That means, according to the creators, that 'increasingly everyday life events take place in the virtual world'. Four years later, in 2003, the same phrase can be used with an opposite meaning: the virtual is being progressively assimilated into everyday life because virtual merges into physical.

(Physical) space as a multiuser environment
With the emergence of the Internet, many authors started to describe cyberspace as a place in which new types of sociability would be developed. However, although there are many virtual multiuser places, urban spaces have become increasingly spaces of displacement. Especially since the development of transportation technologies, people circulate progressively faster across the city space but do not stop to experience the environment or to communicate with each other.

Nevertheless, nomadic communication technologies are restoring urban space with the multiuser feature. That means that urban space has the potential to become again a place.[2] In order to affirm this, it is critical to define what a multiuser environment has become after the cyberspace era. In a summary, MUDs are social places, places that allow communication among people who are not in the same physical place and places that let people inhabit the same (virtual) space even if they are not actually talking with each other.

In a historical perspective, communication has mostly happened in physical space when the speed of travelling and/or circulation was slower and people were able to meet each other while on the move. With the rise of the Internet, communication moved partially to virtual spaces, in which one could experience instantaneous time while sitting in front of the computer. Since the emergence of nomadic technology devices, multiuser environments take place in a hybrid space. Consequently it is possible to communicate with people who are not physically present while moving through space, which is also inhabited by other people. It is exactly the enfolding of contexts that creates the multiuser experience.

There are basically two ways in which cell phones could transform (urban circulation) spaces again into (social) places. One is to enhance communication between people who are close in physical space (referring back to the original meaning of a multiuser environment). The second is increasing communication in the hybrid space, thus changing people's perception of space itself. By bringing distant people into the nearby context, it is critical to question 'what does it mean to be present nowadays?' Of course, the wired telephone had already proportioned the feeling of voice–related presence. However, it is just the mobility that creates the hybrid space. One of the first experiences of enfolding contexts while moving through space was perceived with the Walkman in the 1980s[7] (Hosokawa 1997) but these folded contexts were not connected to two–way communication. Hence the hybrid multiuser space only appears when there is mobility *and* communication involved.

Cell phones and folded contexts

According to Hayles, 'context is becoming enfolded, for there is no longer a homogeneous context for a given spatial area, but rather pockets of different contexts'.[3] For example, someone talking on a cell phone is part of the context of people who share the same spatial area but she is also part of a distant context because she is talking to someone who is spatially remote from her area. So there is a context created by the spatial proximity of people and inside it another context produced by the cell phone. This might be a feature of other media as well, such as the TV or wired telephones. The difference here, however, is precisely the act of moving through space.

This enfolding of contexts reconfigures the way we experience public spaces. They are sometimes not welcomed by people who are against the 'privatization of public space'. However, in many cities around the world people are learning how to live with the 'always–on' situation. This feature is not only related to voice though. 3G cell phones allow the importing of almost any type of information and the injection of it into any situation. Therefore cell phones also expand what the Internet can be. According to Rheingold (Rheingold 2002), even what we understand by the Internet today can completely change owing to mobile technology. This is already happening in Japan where cell phones are sometimes the first Internet connection devices for young people. These kids, when using cell phones to download karaoke or to buy a soda in the vending machine, do not even realize that they are 'using the Internet'.

Cell phones and places

Helsinki and Tokyo can be considered paradigmatic cities regarding the use of cell phones. In Japan, for example, cell phones are used in a completely different way from the rest of the world. The I–mode standard has become a fever and mobile phones have become fashion items, as well as identity objects.

Also common in Japan is the use of interpersonal awareness devices, like the Lovegety, which was not originally connected to cell phones. Likewise, ImaHima –a location–specific application for the Japanese I–mode standard and WAP technology –makes the principle of newsgroups mobile and displays to the user people with the same interests and friends nearby on their mobile phones. Working as a mobile ICQ, each person must give permission before someone else can know automatically where he or she is. ImaHima won the Prix Ars Electronica in the category Net Vision/Net Excellence in 2001.

Similarly, cell phones enabled with GPS are responsible for the development of location–based mobile games. *It's Alive*, a Swedish mobile game developer, created the world's first location–based mobile game, *BotFighthers*. The game takes advantage of mobile positioning and let users play against others in their vicinity by using a standard GSM phone. Each person creates a 'bot' in a website, names it and arms it with guns and shields, similar to a traditional role playing game (and consequently the original MUDs). When their mobile phones are on, the players receive SMS messages about the geographic distance of other players. When they are close enough, they can fight and kill each other remotely, depending on who has more guns and skills. The same company has launched also *X–Fire* and *Supafly*, the first location–based soap opera. These games are an example of how multiuser games formerly played in virtual space can now take place in physical space, taking advantage of the users' mobility.

In every place, however, an important indicative that technology is becoming ubiquitous happens when art starts dealing with these devices and pushes their limits further.

Wop art

Brazilian artist Giselle Beiguelman is a pioneer working with mobile technologies and remote Internet interventions in public spaces in this country. Her piece *Wop Art* (2001) (see http://www.desvirtual. com/wopart /index.htm) connects WAP technology and op art. 'How to deal with an art form conceived to be experienced in between, while doing other things?' is what the user reads on the project home page. The artist's aim was to create a paradoxical situation, since the image in optical art only acquires meaning depending on the viewer's degree of concentration and introspection. On the other hand, generally images conceived for mobile devices do not allow contemplation; they are supposed to be seen while in transit.

The work consists of a series of eight screensavers that can be downloaded to the cell phone. Each one has a different theme: sea, streets, exit/noexit, crowd, difference, egotrip, wysiwyg or x/z=n, and 2beiCode39. Wop Art is a simple example of how cultural content can be disseminated by the use of nomadic technology.

At this point it is interesting to think about some problems and issues that the wide use of cell phones, as well as other nomadic technology devices, might bring. In the cyberspace era, questions about presence, identity and activity were major problems when designing virtual worlds. In a summary, problems were related to how to represent space and the body on the other side of the screen.

Now, what does it mean to be present?

If we do consider our physical space as a multiuser environment, cell phone users (that is, only the ones who possess the right interface) can be viewed as living in an imaginary space. Consequently we could believe that mobile phones withdraw us from physical space, projecting us in the completely imaginary. According to Norman Klein, 'there's no longer need of the screen, because the real world around us has becomes the screen.'[4] Therefore when people talk to each other while on the move, they just walk through space but they are not actually there, becoming walking avatars. This perspective 'generates a culture of tremendous paranoia and isolation. The more we promote an invasion of privacy, the more we make ourselves isolated from the world around us.'[5]

Other questions are related to cell phones as interrupters of social connections, instead of promoting it. The artwork *Social Mobiles* (Crispin Jones and IDEO 2002),[6] award winner at the Japan Media Arts Festival this year, is a critic of this point of view.

Every time a new technology arises, fears and new imaginations are born together. Stories about the fear of travelling in trains come along with the development of the railroad. Also common are questions about the good or bad influences of each new technology. However, some real issues with which content providers should start dealing are:

How to adapt content for a embedded media or a media that is used 'in between'?

We should not take for granted that cell phones enhance social communication, developing modes of cooperation and not isolation.

I believe that some clues on how new mobile technologies are going to evolve will come straight from the arts. The role of the artist has always been to push further the limits of technology, dealing with imaginary spaces and anticipating future.

Bibliography

Donath, J. (1998). 'Identity and Deception in the Virtual Community', in Smith, M. and Kollock, P. (eds.). *Communities in Cyberspace*. London: Routledge, pp. 29–59.

Gibson, W. (2000). *Neuromancer*. New York: Ace Books.

Fortunati, L. (2000). 'Italy: Stereotypes, True and False', in Katz, E. and Aakhus, M. *Perpetual Contact*. Cambridge: Cambridge University Press, pp. 42–62.

Hayles, K. (1999). *How We Became Posthuman*. Chicago: The University of Chicago Press.

Hosokawa, S. (1987). *Der Walkman–Effekt*. Berlin: Merve Verlag Berlin.

Murray, J. (1999). *Hamlet on the Hollodeck*. Cambridge: The MIT Press.

Rheingold, H. (2002). *Smart Mobs*. Cambridge, MA.: Perseus Publishing.

Turkle, S. (1995). *Life on the Screen*. New York: Simon & Schuster, pp. 347 .

3.9 Arch–OS v1.1 (Architecture Operating Systems), Software for Buildings

Mike Phillips and Chris Speed, representing the Arch–OS development team: (B. Aga, P. Anders, M. Beck, G. Bugmann, G. Grinsted, E Miranda, A. Montandon)

Arch–OS v1.1

Arch–OS.com. Software Licence Agreement for Arch–OS (figure 1). Single Building Licence.

PLEASE READ THIS SOFTWARE LICENSE AGREEMENT ('LICENCE') CAREFULLY BEFORE CLICKING THE 'AGREE' BUTTON. BY CLICKING 'AGREE' YOU ARE AGREEING TO BE BOUND BY THE TERMS OF THIS LICENCE. IF YOU DO NOT AGREE TO THE TERMS OF THIS LICENCE, CLICK 'DISAGREE' AND (IF APPLICABLE) RETURN THE ARCH–OS SOFTWARE FOR BUILDINGS FOR DECONSTRUCTION.

1) General

The software, documentation and any building material accompanying this Licence whether on disk, in memory, on any other media or in any other form (real or virtual) (collectively the 'Arch–OS Software') are licensed, not sold, to you by Arch–OS.com ('Arch–OS') for use only under the terms of this Licence and Arch–OS reserves all rights not expressly granted to you. The rights granted herein are limited to Arch–OS's intellectual property rights in the Arch–OS Software and do not include any other patents or intellectual property rights. You own the building into which the Arch–OS Software is installed but Arch–OS retain ownership of the Arch–OS Software itself.

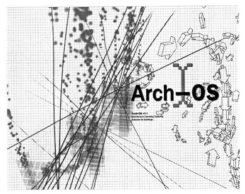

Figure 1: Arch–OS boot screen

2) Permitted license uses and restrictions

a) This Licence allows you to install and use one copy of the Arch–OS Software in a single building at a time. This Licence does not allow the Arch–OS Software to exist in more than one building at a time and you may not make the Arch–OS Software available over a network where it could be used by multiple buildings at the same time. You may make one copy of an Arch–OS building for backup purposes only.

b) Certain components of the Arch–OS Software have been or may be made available by Arch–OS on its Open Source web site http://www.arch–os.com/.

c) Except as and only to the extent expressly permitted in this Licence or by applicable law, you may not copy, decompile, reverse engineer, disassemble, modify or create derivative works of the Arch–OS Software or any part thereof. THE ARCH–OS SOFTWARE IS NOT INTENDED FOR USE IN THE OPERATION OF NUCLEAR FACILITIES OR OTHER BUILDINGS IN WHICH THE FAILURE OF THE ARCH–OS SOFTWARE COULD LEAD TO SOCIAL IRRESPONSIBILITY, ENVIRONMENTAL DAMAGE, SEVERE MENTAL OR PHYSICAL PERSONAL INJURY, OR DEATH.

3) System application

Arch–OS is an 'Operating System' for 'Cybrid' architectures. Cybrids, a term coined by Peter Anders, are 'native to the increasingly mixed reality in which we now live. They integrate physical and cyberspaces within new entities comprising elements both material and virtual. In so doing they marry the affordances of digital media –among them virtual reality, telepresence and online environments –with the grounding stability of matter. In cybrids physical and virtual domains become interdependent: actions in material and virtual spaces mutually affect one another.' Arch–OS, 'software for buildings', has been developed to manifest the social, technological and environmental life of a building and provide artists, engineers and scientists with a unique environment for developing transdisciplinary research and production. Arch–OS has been integrated into the fabric of the University of Plymouth's Portland Square building, which houses the headquarters of the Institute of Digital Art and STAR (Science Technology Arts Research). It has also been commissioned for installation into the three new buildings of the Peninsula Medical School, distributed across the southwest of England. The PMS is a unique 21st century model for the education of medics in a diverse rural peninsula. Arch–OS extends the social and learning communities of these individual and distributed spaces by providing a dynamic networked collective public space.

4) Framework

The Arch–OS provides a framework for 'tele–social navigation' in buildings that are far too complex to understand just by looking at them. Tele–social navigation refers to the feedback loop that exists when the movements of people are modified by environments that are responsive to the interests of the crowd. The Arch–OS project was born out of the desire to explore and illustrate the complexity that defines contemporary buildings. One form of knowledge that the experience of architecture evokes is a social one: the influence of others activities upon our own and a shared understanding of a space. Social navigation, the study of social groups and their influence upon their own environments, provides a dynamic source of data, which transforms the architect's drawings, the brick, steel, glass and fiber–optic infrastructure into a living breathing environment. Arch–OS provides users of buildings with a spatial and temporal consciousness, essentially reprogramming human activity through a heightened social and architectural awareness.

The Arch–OS combines a rich mix of the physical and virtual into a new dynamic architecture. Arch–OS uses embedded technologies to capture audio–visual and raw digital data through a variety of sources which include: the 'Building Management System' (BMS) (which has approximately 2000 sensors in the Portland Square development); digital networks; social interactions; ambient noise levels; and environmental changes. This vibrant data is then manipulated and replayed through audio–visual projection systems and broadcast through streaming Internet and FM radio.

By making the invisible and temporal aspects of a building tangible (see figure 2), Arch–OS creates a rich and dynamic set of opportunities for research, educational and cultural activities, as well as providing a unique and innovative work environment. The Arch–OS takes the notion of 'smart' architecture to a new level of sophistication. A cybrid is an 'intelligent' entity: it interacts, responds and anticipates, and Arch–OS is its nervous system.

5) Systems

There are three system levels to the Arch–OS building:

a) Interface: the construction of the internal media networks and data collection

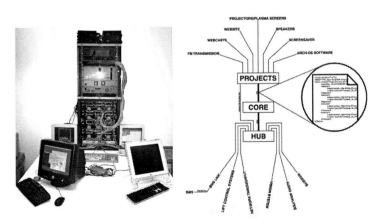

Figure 2: Arch–OS system diagram

devices. The interface (between the physical and the virtual) consists of a dedicated network that transports data from a range of sensors (intelligent cameras that monitor the 'flocking' of people, microphones to monitor ambient sounds, BMS information, network traffic data, lift location and movement) to the 'Core'.

b) Core: the processing and manipulation of the dynamic data generated by the 'interface'. The Core computer systems incorporate a range of interactive multimedia applications (video and audio processors, neural networks, generative media, dynamic visualisation and simulation software) that generate a dynamic 3D sonic model of the building and its activities. This model allows artists, scientists and engineers to manipulate and control the buildings media output which can be broadcast within and between each structure, and out over the Internet. The Core enables the sensing and monitoring of social, spatial and technological interactions such as:

- the movement of people and spaces occupancy can be translated into metaphorical representations such as flocks of birds and many forms of natural phenomena: clouds, waves, buildings being constructed, viruses forming and collapsing

Figure 3: Arch–OS Core model

- temperature can be read and again translated into images and forms, and particularly into lighting systems that modify colour and ambience

- exploring 'Lift Zoning', we are able to develop interesting programming techniques that will make the lifts more intelligent, able to learn user habits and needs and provide a far more intelligent service to the standard dumb lift

c) Projects: the projects enabled by the Arch–OS system are the audio–visual manifestation of the dynamic data processed by the Core.

Figure 4: Arch–OS Sloth–bot

The projects component of Arch–OS is a curated ongoing programme of cultural events, musical performances, installations and exhibitions that take advantage of the unique digital opportunities presented by the building. The institute of Digital Art and Technology (i–DAT) is housed in the centre block of the Portland Square development and will develop, exploit and curate the Arch–OS Core systems to display and disseminate digital works produced by transdisciplinary practitioners. Example projects under development for the Portland Square implementation of Arch–OS include:

i) Sloth–bots: the Sloth–bot (see figure 4) develops the autonomous robotic technology previously incorporated in work such as Donald Rodney's 'Psalms' autonomous wheelchair. These fully mobile robots are spatially aware and sensitive to interactions from passers by and are integrated into the atria furniture. 'Sloth–bots' are large architectural robotic constructions that move almost imperceptibly slowly. Equipped with sensors to keep them from blocking each other, people and sensitive areas, the sloth–bots creep around the atria area reconfiguring the architecture and responding to the flow of people and the movement of each other.

ii) Random Lift Button: Random Lift Buttons have been installed into the new Portland Square Building as a component of Arch–OS. The Random Lift Button project was conceived as an opportunity to exemplify further the role of space at the mercy of time. Certainly in large commercial buildings lifts are implemented to squash space and enable people to move more quickly from one work activity to the next. Lifts become a temporal slippage in the experience of a building as a whole; we skip space and avoid people, places and the opportunity to see the 'whole'. Indeed, corridors and stairwells are recognised as important social spaces within businesses and many more negotiations and affairs occur between office spaces than within them. Just as in hypertext, our choice of destination is provided to us with the minimum of 'journeying'. The Random Lift Button is exactly what it suggests: a button to take you anywhere in a building, thus expanding the space and enabling you to visit spaces that otherwise the economic architectures of today would attempt to hide you from.

iii) Generative Symphony: for the opening of the Portland Square development a generative symphony has been created through the synthesis of human utterances. Hybrid voices are constructed by the system which wail and chant somewhere between the human and the simian. The symphony feeds off the data generated by the Core model and takes advantage of the three–dimensional audio system installed in the building. This consists of a 56–speaker audio mixer that allows sounds to be panned within the three dimensions of the buildings atria to any specified location. When coupled to the vision system specific soundscapes can be 'attached' to individuals walking around the public spaces of the building.

Further planned commissions include:

- the modelling of the microscopic colonisers of our bodies, particulate airborne organisms and residents of our furnishings and appliances exist in great number and diversity

- collaborations with Peter Fend and Ocean Earth

- art science collaborations with the Eden Project

6) Third party acknowledgements
Portions of the Arch–OS Software utilise or include third party software and other copyrighted material. Acknowledgements, licensing terms and disclaimers for such material are contained in the 'online' electronic documentation for the Arch–OS Software and your use of such material is governed by their respective terms.

3.10 (an)Architecture, Eros, Memory: the Naxsmash Project

Christina McPhee

> *We thus enter a universe in which logic does not act as a guarantee of truth; instead truth acts to guarantee the comprehensibility of logic' (a Heideggerian kind of universe, then), harnessing the letter into a dialectic whose very openness is its best guarantee of closure. 'La lettre, ça se lit,' Lacan writes –but this writing is already read, needs no reading from us and is enclosed in pure self–affection . . .With this the prospect of metalanguage collapses, leaving in its stead a problem of imitation and a vision of psychoanalysis as only infinitely prospective and subjunctive science . . . adequately described as pas–tout, its truth elusively and familiarly figured as woman, everywhere and nowhere, not–all.*
>
> (Melville 1996 p. 105)

Subject and place

To begin with, a short story: a phenomenology, of an installation:

At the entrance, a dark space. Within are suspended a forest of dark, long, transparent scrims. It's difficult to avoid touching them. You must negotiate them as you pass through. Inside, lcd projectors arranged in a transverse triangle emit sound and light through scrims, illuminating and interrupting surfaces and image. The sound has recursive structures, so that it seems to propel itself within, around and through the internal spaces generated by the scrims. The music turns itself upside down, inverts itself, slips and falls, and insistently gets back up again as a fugue interaction with the physical space.

Once inside, you obstruct video projection and hear the sound from moving points. Light and image project through your movements and onto your skin. As you enter your reaction and response to the assault of light and sound mediates the presence of the scrims; it activates them as a layering of screens. In reaction to your entrance or non–entrance, to your engagement or non–engagement, to response, the next thing you do or do not do, is performance. Now not only your own screen and audience but you are also the audience for others' bodies–as–screens. There is, within this space, if you want it, an ungrounded experience to take place, even at the same moment that you experience the groundedness produced by the experience of your own, conscious, intentional behaviours and choices. It is, in short, a constructed experience of the Uncanny, or, in Lacanian terms, an eruption of the Real.

The scrims are not empty film, waiting for exposure: there is something upon them. What seems to be on or in them would appear to be still or photographic images but it is not clear to you that they are in fact photographs. They are saturated, with streaks of orange and red in rope–like strands. Here and there you can perceive the image of a woman –or parts of her: her hands, her eyes. It seems to be a figure that's locked up, trapped somehow within or behind the scrims and cannot fully be seen but she gives you the impression that she is not only seeing you, she is looking at you. The videos also seem to have something to do with this woman moving within the narrow, hollow projections of light. Her face and flashes of her body appear and disappear, in layers. The gaze of a double, her tied up hands, her breathing, move through the scrims.

You are aware of the communal aspect of your isolation that arises from your awareness of the other bodies reflecting and refracting among the labyrinths of scrim. There is a seductive quality to the violence of the experience in that you know that it is not merely yourself and the screen engaged in this ebb and flow of light, sound and movement. You experience also, however anonymously, a sense of community with the other disembodied, reflective, severed bodies within the space. You are each of you not only illuminated, spot–lit as it were: you are also comfortingly anonymous in your experience of dis–and re–embodiment, so that you are not only alienated by your experience but you are also, by virtue of the shared aspects of your experience, given permission to enjoy the strangeness, to breathe freely for awhile in *eine Fremde*, a strange land, as Kafka writes of his protagonist K, when K loses control and finds himself breathing for a period of hours in the arms of a woman he does not know (Kafka 1997 p. 38).[1]

Feminine and memoire
The story of this breathing space is Naxsmash, a multimedia performance project now three years into the making.

NAXSmash comes from NAX and subsequent performance works in the series 'Memoires of a Cyborg'. NAX involved rediscovery of a site of childhood violence. The name is shorthand for Lake Nacimiento, a place I had long searched for and finally found. I wanted to go to this lake and make a performance video as a way of getting in touch with the traumatic memory at the site of violence. The video documents an act of breathing as if to contain and release traumatic memory from the site. Memory is the recognition or storage of events; memoire is narration of memory. The video was not memoire because my performance did not tell a story. All I did was practically nothing: an act almost negligent, and subtle, just breathing.

Saving files, I typed 'nacimiento' then 'nascent' then 'nax'. 'X' marked the place but where was it? Nowhere but inside the digital video edits, via erasure and inscription. Smashing the violence through the recovery and digitalising of a violent memory inscribes the memory in a realm that has no location outside the digital object itself. Concealed in a pun, my 'X' factor spliced X as the sign of the feminine inside the media space, as if violated by continuous and limitless edits. I noticed a shift: what had happened to the feminine X, the spot where I was or am, the location of the subject? I was gone, baby, gone. I became witness to my own disappearance. Transposing performance in a new key, in streaming online, in Flash, in installation, in hypertext, I lost track of narrative space. What was there instead? The hallucinatory and decentred aura of the media space was interesting because it was permeated by presence. As if they were there somewhere below or behind the screen wanting to express themselves, decentred subjects moved into the subjunctive mood. In English, we say, 'if we are to go somewhere', if 'you were to come here' –a transactional, formalized ambiguity. The subjunctive mood became a virtual memoire.

It is a bit like old times in those high school nights when 'everybody knows' or better, what if 'if everybody were to know' that there is some girl who is always getting fucked, night after night, behind the bleachers at the school football game. Then what? She is there but we really can't, or won't, see her; she gazes at us in a dematerialized pathos without a story to tell because we are not present to her, to hear her; she is just some girl. I noticed the breathing action by the girl in my video was not 'me'. She was submerged or hidden in the pixels. Violent memory I had released into a cyber spatial transaction but the memoire of that memory was there and not there. Like the raped girl, I could not 'see' myself or 'hear' myself in the performance work. I shifted to a position of working performatively as a cyborg.

Performance atopia
I built online streaming worlds and live installation out of this position: my consciousness was reframed as object inside the media space. From this position I could work powerfully. I witnessed a new fluency in my sound art, digital stills and time–based media. The work was flowing from a decentred subject, i. e. as I had found a formalized way to sound out story, to make music and image without repression, because 'I' was nowhere to be found. In my real life, one of the aftereffects of trauma is silence and hiding. One does not want to be seen because one fears being hurt again. By shifting across the ambiguous space, from the paralysis and silence of memory of rape, into the mediated double of self on and in the screen, I gathered momentum. I now could make work as a series of processional moments out of this traumatised consciousness. Between violence and sublimity and between a subjective presencing and human interactivity, I built cyber spatial narratives as cuts, or smashes, between layers of ambiguous screen. Like pointing to a topology behind, or beyond the screen, the phenomenology of presence generated exactly from atopia, from formal transformations in digital media, none of which imitate anything we can know. There is nothing there

behind the bleachers at the football game. Just code. But it feels real and that is the interesting part. Because it feels real, it becomes active as a topology.

This position, from within the atopic ground, is, paradoxically, negative. It is marked (or stained, to use a favourite Lacanian cliché) by 'X' –amusingly, also the chromosome that, if doubled, produces female sexual beings. With Lacan we can say that it is feminine, as in 'not–all' and 'everywhere'. From the impossible oxymoron of 'woman artist', I have disappeared like the NAX girl into the oxygenated pixels. Inside, my breathing is a consciousness reframed by the edge of the screen. The consciousness is an active viral force field, *pas–tout*' and *partout*. This means that I cut through the spaces between everywhere and nowhere. My artistic work performs as if it is of *jouissance*, the excess, the erotic 'too much' beyond the frame or scrim or screen. Just move along the edges of the images, tracing the change. At the end of the day one is still left with the screen but one has invested itself, and oneself, with memory or, rather, memoires, of what was experienced in the mediated, 'naxsmashed' space.

Topology, the logic of place, meets its end as its beginning in the condition of cyberspace as a single surface twisted into a continuous, meta–temporal, reflexive process, a mobius strip. Lacan describes the move/countermove that always ends at the same no–place:

> *Chest law queue le reel new saurian sincere queue dune impasse de la formaliza-tion...Cite formalization mathématique de la signifance se fait au contraire du sens, j'allais presque dire à contre–sens.*

It is thus that the real is distinguished. The real cannot be inscribed except as an impasse of formalization . . . This mathematical formalization of signification is accomplished against the grain of sense –I very nearly said, a contra sense –the wrong way, by interpretation, absurdly (The real in them of its 'fullness' and 'that which always comes back to the same place.') The apparent invocation of place amounts in fact to the eradication of the notion of place itself ('There is no topology that does not have to be supported by some artifice.') (Melville 1996 p. 106).

The object, the NAX girl, is a shadowy presence in the Naxsmash spaces, both online and in installation. Signals of an entrapped being, she inflects the screen, her motility membrane, like a skin or gut wall, through which she utters a breath of scattered speech. This utterance, suggested through the presence of electronically remixed –and shattered –passages of voice and keyboard, loops back from the point of its origin as an oblique narrative about trauma and violent memory, to return, as its point of origin, as mediated displacements. She cannot be evoked except through the no–place of digital media where she exists nowhere and everywhere, inaccessible and yet full of observable gestures whose significance we invest, or divest, with memoires and desires.

Bibliography

Kafka, Franz (1997). *The Castle*. Trans. J. A. Underwood. London: Penguin Books.

Melville, S. (1996). 'Psychoanalysis and the Place of *Jouissance*', in Seams,. *Art as a Philosophical Context*. Amsterdam: G+B Arts International.

Notes

1. 'Then she started up, K. having remained lost in thought, and began to tug at him like a child: "Come, it's suffocating under here," they embraced, the little body burning in K.'s hands, in a state of oblivion from which K. tried repeatedly yet vainly to extricate himself they rolled several steps, thudded into Klamm's door, then lay in the little puddles of beer and the rest of the rubbish covering the floor. There hours passed, hours of breathing as one, heart beating as one, hours in which K. constantly had the feeling that he had lost his way or wandered farther into a strange land, than anyone before him, a strange land where even the air held no trace of the air at home, where a man must suffocate from the strangeness yet into whose foolish enticements he could do nothing but plunge, on, getting even more lost.' This passage is performed online in my *slipstreamandromeda* www.naxsmash.net/slip/index.html

3.11 Who Plays the Nightingale?

Claudia Westermann

The tender melancholic space constructs itself by obstructing all view; by downs and lowlands ... Silence and solitude have here their home. A bird, which flutters around unsociably, an incomprehensible whirring of unknown creatures, a dove, which coos in the hollow top of a defoliated oak and a nightingale, which has gone astray and moans its woes to the desert, are already sufficient for the setting of the scene.

(Hirschfeld in Böhme 1995)[1]

... there is no psychological theory (in my opinion not even any form of scientific theory), which in the end does not root in self–observation.

(Wiener 1990)[2]

Theoretical systems

'A proof is a finite series of formulas with specific definable properties,' writes Kurt Goedel in his scientific essay 'About Formally Undecidable Propositions of the Principia Mathematica and Related Systems', which was first published in 1931 and later became known as Goedel's Incompleteness Theorem. Goedel proved that a formal system with its defined axioms, contrary to the thesis of that time, cannot decide every question arising in this system. A consistent system must always be incomplete. It can produce more true statements than it can prove by using the axioms of the system itself. The statements become provable only from outside the system but this would mean to widen the system and again one is confronted with a system to which belong more true statements than are provable. Goedel's Theorem had a major impact not only on the science of mathematics but on every science operating with logic conclusions as it implied that every theoretical system must be logically incomplete.

Kurt Goedel's research put an end to the long ongoing attempt to define a general system to solve any given problem. The scientist's dream of being able to develop a universal language as it was formulated, for example, by Leibniz in his 'Dissertatio de Arte Combinatoria', published in 1666, was revived in recent years in the research into Artificial Intelligence, which is seen by many as one of the scientific disasters of recent times.

Technological systems

In contrast to a scientific system –a system that is built on the theoretical idea of logic –a technological system as an application and materialization of the scientific system must be provable. It is limited to a range of only provable statements.

Progress, in fact, is not a growth of ideas but a success in a reduction of ideas to a limited range of only provable statements.

The system city

In 'L'invention du quotiden. Arts de faire', which was published in French in the year 1980, Michel de Certeau describes the city as an alliance of 'fact' and 'concept'; thus he defines city as an alliance of a provable and a non–provable system. His differentiation in between location and space defines location as something that can be planned and space as something that is created by actions, movement (de Certeau 1988 p. 183). Thus the concept city implies the dimension of time and can be linked to another important term, the atmosphere.

Gernot Böhme defines the term 'atmosphere' in the following way. 'One sees elements in their arrangement, things, which refer to each other; one sees situations ... Situations concretise themselves only from case to case before the background of a world. However one does not see the world. But what is this whole, then, into which every individual [thing] is incorporated and which, depending on attentiveness and analysis, one can point out? We name this primary and, in a certain sense, basic object of perception the atmosphere ... so important is it that every individual [thing] is coloured, in a way, by the atmosphere' (Böhme 1995 pp. 94–5).[3]

Blue room

Perhaps once in time a blue room will go travelling around the world, temporarily occupying urban space and then leaving that space behind again but with a remembrance of something blue, connected to all the other blue spaces the room will have left behind ...

Inside entirely blue, light entering only between the ceiling and the walls. Four speakers transmitted a sine wave sound, which by means of superimposition extinguished itself in regular sequences. For one day the blue room coexisted in suspended animation with the church behind it, taking possession of the plaza, much as the sound of a church bell creates a sphere of influence. The relationship of the blue room and the church is not one of an opposition (that would be too strong), however, positioned symmetrically to the church on its own plaza the relationship was obviously a challenging one. Blue room provided for an alternative to the historically sacral space, a de–institutionalized, non–historical sacral space, thus possibly not more than a meditative space

The unprovable city

In the passage quoted in the beginning of this paper from *The Theory of Garden Art* by C.C.L. Hirschfeld, the author's attempt seems to be to make the unprovable plannable, to actually develop a universal language for a concept. The incompleteness of such system is most apparent. Atmosphere is a theoretical system; it cannot be proved.

In the field of architecture the generalist approach is not unfamiliar. In fact, the need to operate in theoretical as well as in technological systems perhaps makes a more generalist approach necessary and the attempt to develop a language that is valid for both systems is understandable. Le Corbusier's development, for example, of the modulor can be seen in this context.

A similar attempt is Christopher Alexander's development of a 'pattern language' whereby he clearly states its incompleteness and asks for individual pattern languages with common patterns to be shared. In contrast to Le Corbusier, he addresses neither the architect nor designer but rather asks for cities to be built 'alive' by the people inhabiting them. In this way he is very close to Joseph Beuy's statement that 'everyone is an artist' and the 'pattern language' is possibly, too, much more of a statement than being meant to provide actual means for construction.

For de Certeau the concept city is highly dependent on the notion of movement and it can be perceived today that the public spaces of a city actually are mainly spaces of transition. In contrast to de Certeau, who saw mainly beauty in the movements within the city, one perceives these movements today as being a danger as well, turning public spaces mainly into spaces of transition and much less into spaces of communication, a development that has become faster since mobile technology entered city spaces. It seems that people transit the public spaces in 'bubbles' of private spaces. The potential for communication has decreased.

Zone_01
The sound installation zone_01 is designed to be realized in public space. In the course of progressing technification, public space is losing its significance as space for communication. Public squares are experienced as interruptions within paths of transition. zone_01 simulates communicative processes and translates them into sound. The movement–oriented use of the location is transformed into a sonorous simulation of the communicative function of such a public square and leads to changes within the system of sound.

The virtual city
The virtual space has many things in common with actual urban space. A society, which is a theoretical non–provable system, operates through an alliance of a provable and a non–provable system. The Internet as the technological application of a scientific idea is the provable system. The network system is the scientific non–provable one. An actually working open society in a network can be mainly seen as a simulation of a scientific space. For the construction of communication

spaces in the fluid environment of the Internet, the virtual architect faces a shift towards being an ambiguous figure: voyeur, object, creator as well as material.

'I am ONE. I walk to be ALWAYS THE LAST in my sequence. My memories are operators to my dreams,' says the girl in the virtual garden and continues.

The rectangular room is painted in bright white. No window or door disturbs the continuity of the walls. However, there must be openings in the ceiling since natural light enters the room from above, along the walls, as funnel–shaped rays.

A long shelf, approximately 1.2 meters in height –a question of perspective –and painted white as well, seems to belong to the room. It is placed within a distance from every wall. In its exposed position it fails to correspond to a wall's parallel. Photographs are placed on the shelf, organized in a line but in unequal distances from each other –a moderate number of portraits of females in different ages framed in white. The line of photographs starts at the left narrow side of the shelf and continues to its middle. She sits next to them, just beside the intersection with the second half of the shelf, and does not move, dressed in black. Across the room her eyes seem to focus on an unspecified point on the sidewall.

At a secure distance from the shelf, a heavy armchair establishes a living room culture in brown. On the chair someone has left a scarf and a book, open with the pages turned downwards.

The light fades. It is evening.

Bibliography

Alexander, Christopher (1979). *The Timeless Way of Building*. New York: Oxford University Press.

Alexander, Christopher (1977). *A Pattern Language*. New York: Oxford University Press.

Böhme, G. (1995). *Atmosphäre*. edition suhrkamp.

de Certeau, M. (1988). *Kunst des Handelns*. Berlin: Merve–Verlag.

Le Corbusier, (1948; in German in the 7th edition, 1998). *Der Modulor*. Frankfurt: DVA–Verlag.

Dawson, John W. (1999). *Kurt Gödel –Leben und Werk*. Wien, New York: Springer.

Gödel, K. (1931). *Über formal unentscheidbare Sätze der Principia Mathematica und verwandter Systeme*.Wien: (a text version in English is available on the Internet at http://home.ddc.net/ygg/etext/godel/godel3.htm

Hofstadter, D.R. (1979). *Gödel, Escher, Bach: An Eternal Golden Braid*. New York: Basic Books.

Lootsma, B. (1998). 'Public Space in Transition', in *Daidalos*, no. 67, Gütersloh: Bertelsmann Fachzeitschriften GmbH, pp. 116–123.

Wiener, Oswald. (1990). *Probleme der künstlichen Intelligenz (Problems of Artificial Intelligence)*. Berlin: Merve–Verlag.

Notes

1. Hirschfeld, *Theory of Garden Art*, quoted in Böhme, *Atmosphäre*, p. 31. Translation by the author, original citation in German: 'Die sanftmelancholische Gegend bildet sich durch Versperrung aller Aussicht; durch Tiefen und Niederungen . . . Die Stille und Einsamkeit haben hier Heimat. Ein Vogel, der ungesellig herumflattert, ein unverständliches Geschwirre unbekannter Geschöpfe, eine Hohltaube, die in dem hohlen Wipfel einer entlaubten Eiche girrt, und eine verirrte Nachtigall, die ihre Leiden der Einöde klagt –sind zur Ausstaffierung der Szene schon hinreichend.'
2. Wiener, *Problems of Artificial Intelligence*, p. 56. Translation by the author; original citation in German: '. . . es gibt keine psychologische Theorie (ich meine sogar, überhaupt keine wissenschaftliche Theorie), die nicht letzten Endes in der Selbstbeobachtung wurzelt.'
3. Translation by the author; original citation in German: 'Man sieht Dinge in ihrem Arrangement, Dinge, die aufeinander verweisen, man sieht Situationen . . . Situationen konkretisieren sich nur je von Fall zu Fall auf dem Hintergrund einer Welt. Allerdings die Welt sieht man nicht: Was aber ist dann dieses Ganze, in das alles Einzelne, das man dann je nach Aufmerksamkeit und Analyse daraus hervorheben kann, eingebettet ist? Wir nennen diesen primären und in gewisser Weise grundlegenden Gegenstand der Wahrnehmung die Atmosphäre . . .Wichtig ist, dass dann jedes Einzelne gewissermaßen voen der Atmosphäre getönt ist.'

3.12 Breeding, Feeding, Leeching

Shaun Murray

To *dissect* a space and break it down into bite–sized chunks, and then isolate each one from its environment, then try and put the pieces back together again and fragment the genealogy of an architectures development.

Within these sets of relationships I will *study* and *dissect* three architectural experiments governed entirely by theoretical principles, which permits the exploration of the consequences of these principles in a given situation. My work has long explored these territories, proceeding along its own trajectories from *disturbed territories* to *reflexive architectures*, to my present explorations of the idea of Breeding Feeding Leaching environments, suggesting an architecture that is 'altered'. Spaces are built up of a series of *characters*; these characters can function independently or in a group. Each character has its own sets of rules in space. Some are determined by time and duration, others by soft and hard edges, whilst others are determined by elemental levels, e.g. light levels, temperature.

Breeding

The process of producing spaces and objects by hybridisation, inbreeding or other methods of reproduction. To produce or be produced; generate; a group of characters within a space. To produce and maintain new or improved strains of object or space–time relationships in an architectural context. To breed trouble. A space passing through another to reproduce fragments of the space and play back the fragmented space at a later time; to experience or perceive part of a space as if it has occurred before. To play a space back on itself as a confirmation or engagement with an actual event occurring.

Figure 1: 'How can Architectures Breed. Feed. Leech?' In large part, with the developments in technology and biological implications, with the alchemical grafting into the architectural, and so on

The term *breed* suggests reproduction and consequently a notion of modification as part of a breed; this in turn has the potential to provide an 'altered' space. Its introduction into architectural discourse marks a milestone in the transition of architecture from the passive, manufacture and inevitability to choice, activity, life and, eventually, consciousness. A space can breed from the interaction of two characters from separate spaces. Characters in space contain sub–characters, which are continually 'fluctuating'. A single fluctuation or a combination of them may become so powerful, as a result of positive feedback, that it shatters pre–existing organisations. At this moment there is a bifurcation point –it is inherently impossible to determine whether the characters will disintegrate into 'chaos' or leap to a new, more differentiated higher level of organisation. This is called a dissipative structure.

Feeding

To provide what is necessary for the existence or development of an object: to feed one's imagination; to gratify; satisfy. To supply an architectural object with necessary systems for its operation, or of such materials; to flow or move forwards into a vessel, etc.; the process of supplying a space with a material or fuel. A mechanism that supplies material or fuel or controls the rate of advance of an architectural environment. To provide a space working on the very edges of form, moving with greased abandon between categories such as fibrillation, dissection and degradation.

For a space to feed it needs to have a *defence mechanism*. This would involve characters of one space initially to inhabit other characters in another space whilst allowing the other characters to function as normal. During this process the feeding characters will intuitively put out feelers to examine the environment's strengths and weaknesses before the space becomes 'altered'.

What defence mechanism does the altered space have?

1) spillage in existing space and to decay and strangulate aspects of the other character

2) to have a dependence on existing environment until found out characters of altered space to feed

3) release from the existing environment and acting in suspension (playing dead) while still investigating 'altered' space

4) reflexive impulses: playing out spatial dichotomy of investigating 'spatial swapping' with algorithmic twitching

To *alter* space is to create a hybrid environment through mutation and adaptation. Once the initiation procedure has occurred the following actions are followed:

- feed

- propagate

- grow

This process of adaptation and mutation generates an 'altered' space, which seeks and destroys failing environments. The space sheds its surfaces, exfoliating the logistics of the 'altered' space.

As the 'altered' space develops it engages on a more physical level and duplicates the existing characters of the space. This dysfunctional replication of the existing is embroiled in disseminating acts of characters.

Leeching
An architecture that clings to or preys on another architecture. An archaic word for physician; 'leechcraft'. To cling or adhere persistently to something. To remove or be removed from a substance by a percolating liquid. A porous vessel for leeching, perhaps from Old English leccan, to wate. Related to leak.

Leeching aims to examine new alternatives to other spaces' characters. By reusing existing characters from other spaces to accommodate altered programmes of characters, strategies would be developed for investigations which insinuated themselves into existing environments, contaminating and infecting the infrastructure of the altered space. 'Host' environments would be revitalised by a variety of parasitic interventions, thus avoiding degradation.

What results is a prosthetic of additions, of invasions and infections. The host environment is not merely disregarded and demolished but used as a provocative point of departure for new altered spaces.

Antithetical programmes for the characters would be developed to provoke a pragmatic clash within the existing use of the host. Strategies for the development of superimposed yet contiguous programmes will increase fragmentation of the existing space.

Space will become entered apprentices in *character space*, to generate an environment of attraction and repulsion. One entered apprentice would be attracting its enemy's enemy with [s]light movements and suggestive repetition. To explore how objects can engage in sharing space while their edges are communicating to each other with the invisible forces of choreography. I want the work to powerfully suggest a weightless space, like tranquillity and silence. Events act as a catalytic moment in time when doors are left open and the conclusion of the piece is left to linger.

It becomes an architecture in search of a physical form but derived and controlled by the physical stimuli of this unique local environment. The work becomes the intermeshing of differentiated local stimuli in various natural environments as control factors for the construction of architectural environments.

The following series of theoretical and experimental projects explores the limits of architecture using the defined concepts of Breeding. Feeding. Leeching. Each project is set up as a network of events to allow for the capacity for hybridisation and new 'altered identities'. These altered identities are subject to change over time.

Archulus flood structure
Aldeburgh coastline, Suffolk, England, 2000

The coastline of Aldeburgh is a delicately balanced interface between man, sea and river, subjecting itself to many geo–physical processes. These processes operate over a range of timescales, from short–term fluctuations to long–term changes over thousands of years. 'Archulus' is a consequence of recurring floods on the east coast of Suffolk, which has lost more than a metre a year for the last 400 years. The project straddles this transitional interface and explicitly injects and infuses its own agendas. The movement of waves, tidal imbalances and currents shape an ever–changing Archulus profile.

A model is constructed using a set of derived values from the landscape (temperature, humidity, salinity levels) of the real world and from data and processes of the virtual world. Also, from numerous techniques of capturing the real and casting it into the virtual, the model is fed time–based data and the form becomes animate, the architecture *vacillating*.

The shields act as a giant loom as a reaction to the flood tidal conditions which are inflicted upon this coastline. The act of an environmental and climatic condition has a reaction with the advancement of the fibrillated object. The advancement of the shields acts as a giant loom and generates a woven space behind

Figure 3: Void Scan showing the turbulence voids scanning and mapping the surface topology of the seabed. Dark spaces are recovered, abstracted and observed. This will become a study of an artificial and renegade ecology, whereby the object as organism virtually creates a

Figure 2:

new ecology of an invisible framework. The processes of archaeology and architecture are used to strive for recovery of the old settlements and roman forts and lead to discovery and insight. It will be achieved at times graphically and be exhibited digitally inside the flood structure; at other times a turbulence void will suddenly exposing itself at the surface of the water and freeze part of the animated structure with disconcerting and sometimes devastating effects. Cathode ray scanning of a surface might be translated to a numeric–command machine in order to apply complex calculus derivatives directly to material surfaces.

Figure 3. Darksoul Darksoul Collective, 2001

The work is a piece of sculpture –with no dialogue, a fixation of characters in space; you become the 'entered apprentice' on this non–linear narrative. With surprise of entry and exit, intense detail of pieces are built on a clear purposefulness of the characters. It is a huge motile sculpture, working on the very edges of form, moving with greased abandon between categories such as installation art, rock videos and contemporary dance.

'DEEP in the dark depths there are dragons, and their fire is a beam of red light. Like military sniperscopes, the dragon's red searchlights allow it to see without being seen, except by other dragons equipped to pick out their secret signals. The red light allows them to hunt unseen and find prospective mates without alerting their own enemies. This serves many purposes: to attract a mate, seek out or lure prey closer or frighten off potential attackers. A bright flash of light can blind an approaching predator used to dim light, perhaps for long enough to make a getaway. Things only emit light when something does something, when there is an interaction between dragons. But just one interaction –a dragon blundering into another, for instance, can set off a chain of flashes and twinkles.

'A dragon looking upwards could see a shadowy silhouettes of other characters moving overhead against the dim light above. Some characters can make themselves invisible by counter–illumination, giving out light of matching intensity along their bellies.'

Figure 4: These images show the interiority of the *Collective Voids* as they bask in dark spaces of the sea. These dragons' fire is a beam of red light that shines from a lamp beneath each keel. Like military sniperscopes, the dragon's red searchlights allow it to see without being seen, except by other dragons with lights equipped to

Figure 4

pick out their secret signals. The red light allows them to hunt unseen archaeology and find prospective territories. They only emit light when something does something, when there is an interaction between organisms. But just one interaction –a turbulence void blundering into another, for instance can set off a chain of flashes and twinkles.

Portrayals of degradation
London 2003

Portrayals of degradation was developed to show how the landscape develops its own language and environment of invention and by turning in on itself makes the 'plot' an intimacy of interplay of creation and intrigue. With this controlled and analytical judgement of the sites past and present, an architecture can be intrinsically positioned and will interplay with the forces of what is already there and be played back on itself.

4. The Text
4.1 The Potential of Electronic Textuality

Dene Grigar

As a point of reference, digital literature is a print–based literary work[1] digitalised for the web and standalone technologies like a CD–ROM. Work in this area generated from the fields like Humanities Computing is sustained through numerous conferences and publications, such as the Joint International Conference of the Association for Literary and Linguistic Computing and the Association for Computers and the Humanities (ALLC–ACH), Human IT and Computers and the Humanities and has been underway since the late 1980s. Examples of digital literature would include a copy of Homer's Odyssey found at The Perseus Project[2] and an electronic edition of Emily Dickenson's 'manuscripts'.[3] Digital literature is not to be confused with electronic literature, an emergent literary form and academic field whose genesis can be traced to the 1989 work by Michael Joyce, afternoon: a story. As a form it can be defined as literary works created by computers for the electronic medium; as a field, it is building through online journals, like the Iowa Review Web and Poems That Go, and organizations like the Electronic Literature Organization (ELO).

The rationale for incorporating approaches to textuality associated with the electronic medium is simple. Because the computer was originally built for calculating numbers, its utilization for display and preservation of or communication with words, images and sound was ancillary to its primary development as a tool for computation. Thus these uses of the computer, coming much later in the field of computing, suggest that they have been executed upon a structure alien to them and for an audience not fully prepared for the kind of literacy practices required to understand them. Yet despite this situation, their augmentation, extension and mediation by computers into the computer environment have flourished and have resulted in what Anna Gunder calls a 'media convergence' (Gunder 2002 p. 82) and articulated as 'remediation' by Jay David Bolter (Bolter and Grusin 2000 p. viii) –that is, present always in 'acknowledged or unacknowledged ways' (ibid p. 47) of older, more familiar treatment of display, preservation and communication rather than a move to rethink and create anew. What many scholars and artists involved in digital and electronic literature struggle with is how much to remediate old forms, structures and landscapes in order for others to understand and make sense the literary work created for the computer environment.

Put more succinctly, print–based literature digitalised for the web has in many cases merely replicated old forms rather than created new ones. Even those that have taken advantage of the hypertext, as in texts found at The Perseus Digital

Library, have invoked print–based conventions in computer–based environments. That said, this paper is not an attempt to find fault with current modes of display, preservation and communication but rather seeks to explain why works have followed this method and to offer a perspective that may help to move future work beyond print–based conventions and into computer–based potentiality.

The legacy of print seen in digital literature can be explained from numerous theoretical bases, like Bolter and Grusin's theory of 'remediation,' mentioned above. But certainly three that offer unique perspectives for innovating print–based texts for the electronic medium are translation studies, rhetoric and electronic textuality.

Literary translation, an activity enacted on a written or print–based text, entails a careful study of relationships. In translating literary texts, the translator must look at five different relationships, relationships best described as 'print–textual' ones: the relationship between 1) words, 2) the word and its philological and etymological background, 3) the word and its cultural ambiance, 4) the word and its historical tradition and, lastly 5) the word and its context within a text. In fact, a good translation 'shies away from a statements of interpretation' and instead is 'directed toward the discovery of relationships in a text' (Biguenet and Schulte 1989 p. xii).

The first key for the innovation of print–based texts digitalised for the electronic medium may lie in the word text.

Definitions of text held by literary scholars, as Mats Dahlstrom reports in 'When Is a Webtext?', do not comfortably fit or reconcile easily with current notions of computer–generated and mediated texts. Text, for print–oriented, literary scholars, 'has been the object of metaphorisation and abstraction to the degree that it increasingly denotes signified meaning rather than signifying expressions' (Dahlstrom 2002 p. 141). This approach has led to analyses of the theories about text rather than analyses of a text itself.

Others in the humanities disagree with this approach. G. T. Tanselle's A Rationale of Textual Criticism makes it clear that textual critics treat text differently when he argues that 'all works of art have texts' and texts are 'arrangements of elements' (Tanselle 1989 p. 21). He is seconded by Peter Shillingsburg, who says that 'a text is the actual order of words and punctuation as contained in any one physical form, such as manuscript, proof, or book' (Shillingsburg not in bibliography p. 46). Gunder takes the definition beyond print when she defines text as 'the representation of [a work], through which the work has to be experienced' (Gunder 2002 p. 86). Still others, like Espen Aarseth, coming out of the field of humanities infomatics, define text as 'a whole range of phenomena, from short poems to complex computer programs and databases' (Aarseth 1997 p. 20), as well as 'any object with the primary function to relay verbal information . . . that does not operate independently of some material medium, [medium that in effect] influences its behavior (ibid p. 1). And finally, Katherine Hayles, working in the field of electronic literature, includes in her definition of text images and sound, as well as other 'signifying elements' (Hayles 2002 p. 20). As Tanselle, Shillingsburg, Gunder, Aarseth and Hayles all demonstrate, text can be and is viewed as possessing a material aspect, a physicality that, more than likely, should not be ignored in analysis. To be honest, the conflict boils down to two approaches to textuality that have dominated the humanities in the late 20th century: textological and textonomical.[4] The former, a common approach in postmodern literary theory, disallows the materiality of texts; the latter, common to textual studies, embraces materiality but does not recognise text as art. For both, no matter what the thinking is towards its material aspect, text is merely code.

What makes translation theory a compelling one for moving digital literature beyond print–based approaches is its conscious attention to associations and artistic human endeavours.[5] Viewing the digitalisation of print for the electronic medium this way allows for the consideration all aspects of the object, from the relationships it creates to its sensual qualities –and for the resulting artifact to stand as an artistic work on its own merit. It is very difficult to look at Michael Denner's work with Russian poetry on the web, From the Ends to the Beginnings, and not view it as an artistic achievement in the field of digital literature, so rich is his use of the medium for his text.

The second key for the innovation of print–based texts digitalised for the electronic medium may also lie in the word medium.

Medium refers to the material on which 'a text is recorded' or rather the material 'that carries the signs representing the text' (Gunder 2002 p. 99) or message. Ironically, the emphasis on transforming text, whether it occurs between languages as in translation or between media as in digitalising literature, has been placed on text. Considerations regarding the medium where that text resides have largely gone ignored by some of the very scholars who see the material aspect of text. In other words, for many, the act of moving text from medium to medium has merely been treated as a matter of transportation rather than any kind of transformation. Thus works taken from a page of a book and moved to a computer screen privilege the five print–textual relationships at the expense of any enriched by the computer medium.

What is meant by 'enriched' is that translating a literary work between media suggests these relationships are augmented by three others related to medium and/or the computer: 6) between the story and its medium of articulation –that is, what does the source medium offer the story that helps to construct meaning that the target medium cannot provide, and vice versa, 7) between the reader and the media disseminating the literary work or the assumptions the reader brings to the medium that affect the way she or he constructs meaning from the story, and 8) between the reader and the work itself, particularly interactive works that require a reader's intervention to move the narrative forward.

These three relationships signal a rhetorical perspective, or 'situation' that should also be taken into account when moving text online. In particular, they demand that the creator of a text address her or his audience in the very re–creation of the work so that the reader engages in the work and responds to it in some particular way. This type of rhetorical approach is referred to as 'readerly' and texts that involve this approach, readerly texts,[6] are texts that take into account sound rhetorical strategies relating to arrangement and structure and that attend to rhetorical situations –that is, those situations 'that call for functional uses of discourse to adjust people, objects, events, relations, and thoughts' (Hauser 1986 p. 33). Thus if the reason print–based texts are digitalised for the web or for use with a CD–ROM is for the purpose of making that literature available to a broad audience or for preserving that literature for enduring access for a future undetermined audience, or even communicating an idea in another format, then, rhetorically speaking, when digitalising literature for the web, rhetorical issues relating to 'audience' must also be addressed –a notion that takes scholars squarely into the realm of rhetoric. In sum, at the heart of digital literature projects are issues relating to producing a good translation with a sound rhetorical approach.

But basic assumptions about print textuality many times impede an understanding of the potential of the electronic medium and colour perceptions of the narrative experience. These include immersion, kinesthetic but trivial reading practices, the presence of conflict and tension, the types of time utilised in the work, the containment of space of the text, the personal relationship with a text a reader may experience and specific sensory perceptions required to engage with a text. Solutions may be found by turning attention to the specific media in which the narrative is experienced.[7] In the case of digital literature, the textuality is electronic.

The third key for the innovation of print–based texts digitalised for the electronic medium may lie in the word textuality.

As Katherine Hayles points out, 'electronic textuality' involves multiple senses and takes into account 'other signifying components of electronic texts, including sound, animation, motion, video, kinesthetic involvement and software functionality, among others' (Hayles 2002 p. 20). Works of literature created for the electronic medium –whether that work is native to computers like electronic literature or translated to it as in digital literature –that take into account these qualities of electronic textuality would involve sound, images and kinesthetic activity along with the text and would appeal to senses different from those assumed in the print environment. And indeed, Denner's website makes good use of sound as a way of engaging his audience. Unfortunately, mowever, that is uch digital literature mirrors its print iteration. The argument for media purity, as Sven Bikerts makes throughout his book The Gutenburg Elegies: The Fate of Reading in an Electronic Age, is a specious one if one considers that he thinks nothing of the fact that Homer's epics are printed when, in fact, they were intended originally to be recited orally. That they now appear online is but a third iteration of the original, having gone first through writing, to print and finally to the electronic medium.

In the Phaedrus Plato laments the shift from orality to writing. His stated fear was that the new medium of the page would render ideas 'dumb' (Phaedrus, 275d–e). What he meant was that words would be disembodied from the speaker and so would fall silent, left open for any interpretation –particularly, wrong ones by the misinformed and uneducated –without the philosopher present to provide the truth behind the words, as would happen in a lecture or speech. In essence, Plato was acutely aware that the shift in medium was a significant one and entailed a change in how the text, for him as ephemeral as the wind, would be experienced and understood when it was chiselled in the medium of the proverbial stone. It is no surprise that the conventions that came later, such as divisions into chapters and paragraphs, placement spaces between words and even the various critical approaches, are all attempts to display, preserve and communicate better the ideas contained in a text and can be seen as contributing to the further refinement of textual practices.[8]

In much of the same way a shift from print–based to the electronic medium signals the need for a new sensibility, a new treatment of text, new conventions. What these innovations turn out to be in the final analysis is anyone's guess at this early stage of computer technology but Haley may be right in suggesting that creators of any kind of computer–based literature embrace electronic textuality, for as this approach –along with translation theory and rhetoric –suggests, utilizing the relationships and associations solely connected with print does not take full advantage of those potentially available for development by the creator of digital texts or engagement by the audience with those texts. What is clear is that they need to be augmented by those derived from the electronic medium.

Bibliography

Aarseth, Espen (1997). *Cybertext: Perspectives on Ergodic Literature*. Baltimore: Johns Hopkins University.

Baron, Naomi (2000). *From Alphabet to Email: How Written English Evolved and Where It Is Headed*. London: Routledge Press.

Biguenet, John and Schulte, Rainer (1989). 'Introduction', *The Craft of Translation*. John Biguenet and Rainer Schulte (eds.). Chicago: University of Chicago Press.

Bikerts, Sven. *The Gutenberg Elegies: The Fate of Reading in an Electronic Age*. NY: Fawcett Columbine, 1994.

Bolter, Jay David and Grusin, Richard (2000). *Remediation: Understanding New Media*. Cambridge: MIT Press.

Dahlstrom, Mats (2002). 'When is a Webtext?'. *Text Technology: The Journal of Computer Text Processing*, vol. 11, no. 1, pp. 139–61.

Denner, Michael (2003). *From the Ends to the Beginnings: A Bilingual Web Anthology of Russian Verse*. March. www.max.mmlc.northwestern.edu/~mdenner/Demo/project.htm. Accessed 17 July 2003.

The Dickinsen Electronic Archives (1999). Martha Nell Smith, Ellen Louise Hart, Marta Werner and Lara Vetter (eds.). www.iath.virginia.edu/dickinson/. Accessed 17 July 2003.

Friedrich, Hugo (1992). 'On the Art of Translation.' *Theories of Translation*. John Biguenet and Rainer Schulte (eds.). Chicago: University of Chicago Press, pp. 11–16.

Gambier, Yves and Gottlieb, Henrik (2001). *(Multi) media translation:Concepts, Practices, and Research*. Amsterdam: Benjamins.

Gunder, Anna (2002). 'Forming the Text, Performing the Work –Aspects of Media, Navigation, and Linking'. *Human IT*, 2–3, pp. 81–169.

Hauser, Gerald (1986). *Introduction to Rhetorical Theory*. Prospect Heights, IL: Waveland Press.

Hayles, N. Katherine (2002). *Writing Machines*. Cambridge, MA: MIT Press.

Joyce, Michael (1989). *afternoon: a story*. Watertown, MA: Eastgate Systems.

The Perseus Digital Library. Gregory Crane (ed.). March 1997. www.perseus.tufts.edu. Accessed 17 July 2003.

Schillingsburg, Peter L. (1991). "Text as Matter, Concept, and Action." *Studies in Bibliography* 44: 43–83.

Tanselle, G. T. (1998) *Literature and Artifacts*. Charlottesville: The Bibliographical Society of the University of Virginia.

Tanselle, G. T. (1989). *A Rationale of Textual Criticism*. Philadelphia: University of Pennsylvania Press.

Notes

1. Defined by Anna Gunder by as an 'abstract artistic entity' of which a text is a component (Gunder 2002 p. 86).
2. The Perseus Digital Library is 'an evolving digital library, engineering interactions through time, space, and language. Our primary goal is to bring a wide range of source materials to as large an audience as possible. We anticipate that greater accessibility to the sources for the study of the humanities will strengthen the quality of questions, lead to new avenues of research, and connect more people through the connection of ideas.' It is interesting to note that it began as a repository for all ancient Greek works, including both literature and art. From it origins it has expanded to Latin works, and from its description, it looks like it is expanding further outside the Classics and subsequently will reach a larger, more diverse audience.
3. Appeals for electronic editions grew in the mid to late 1990s as exemplified by those like the Dickinsen Electronic Archives.
4. See Dahlstrom. pp. 139–61. He cites the work of Espen Aarseth from his book *Cybertext: Perspectives of Ergodic Literature*, pp. 15–17.
5. Hugo Friedrich traces the origins of western literary translation to the Roman period, making a strong case for the idea that translation has been a recognised art form beginning with the ancient period. See 'On the Art of Translation,' pp. 11–16.
6. Readerly texts sit in opposition to 'writerly texts' or texts written for the author with little to no attention paid to the audience's understanding or enjoyment of the work.
7. Hayles addresses 'media–specific analysis' in several of her works, most recently in *Writing Machines* (2002 pp. 29–34).
8. For an excellent discussion concerning the development of writing conventions from its roots to the present time, see Naomi Baron's *From Alphabet to Email: How Written English Evolved and Where It Is Headed*, especially pp. 27–47.

4.2 Art and Information

David Topping

And the world itself, certainly the social world, has always relied on its appointed collectors. Civilization could not exist without tax collectors and gatherers of information, harvesters and hoarders, census takers and recruitment officers, rent collectors, ticket collectors, refuse collectors, undertakers . . .

(Elsner 1994 p. 2)

While seeking to find a way of describing my own art practice I began to call what I do Information Art. This is within a wider study considering the use of the word information as a useful and congruent term to describe an area of artistic practice and to explore its historical and current usage. After briefly surfacing at the end of the 1960s, and with one or two exceptions, the word has been notable by its absence. This paper details some of the work done with standardised forms such as loan applications and personnel and job application forms as well as considering the role of artists as information gatherers and processors. A statement for the Tate Britain's exhibition *Intelligence: New British Art 2000* explains, 'Many artists today can be seen as intelligence agents at large in society, gathering, sifting and transforming the raw data of our life, critically examining our environment, the way we live and our relations with each other' (Button 2000).

Three artworks that I wanted to explore in this context were Maya Lin's *Vietnam Veterans Memorial*, Tracy Emim's *Everyone I Have Ever Slept With (1963–1995)* and Gerhard Richter's *Arrest 2, 1988, from the 18 October 1977* series of paintings about the Baarder–Meinhof Group. Though it's possible to see the choice of these works as arbitrary, they are able to provide a pathway that I hope to illuminate within this study. On one level both Emin and Lin's work contain obvious elements of information. They both use text and both are lists, a structured form of information. Emin's *Everyone I Have Ever Slept With (1963–1995)*, a blue tent appliquéd with the names of family members, sexual partners and the foetus she aborted, can be considered in two important ways. Firstly she memorialises her past in a way that parallels the more obvious memorial aspects of Richter and Lin and, secondly, possibly like all good art, it is an exercise in self-regulated integrity. As art they both lose much of their authority if we don't believe them to be both true and, more importantly, complete. In this there is an almost train-spotting-like abhorrence for the inventing of fraudulent entries. Michael Landy's *Break Down*, his destruction of all 7,006 objects he owned, also loses much of it power if we aren't convinced that everything, no matter the commercial value –and he included his Saab 900 turbo car and artworks by Damien Hirst, Gary Hume and Emin –has been destroyed. Though seen by him as a response to consumerism, it is also possible to view his actions as both an attempt to make sense of the world by classification: everything was computer catalogued prior to the event which was the product of the kind of disorder that often externalises itself as autism or obsessive–compulsive personality disorder. This internalisation by Emin and Landy needs to be set in context with the external work by Richter, Lin and other 'memorial' artists such as Rachel Whiteread. Emin uses her personal experience to make highly confessional works, continually referencing her history in an ongoing documentary. Her exhibition *I Need Art Like I Need God* is an extension of *Everyone I Have Ever Slept With (1963–1995)*. Her life, pieces of notes, video, photographs. etc. were exhibited as works.

Standardised forms are implements of collection, used to structure information and, more specifically, either simplify or exclude elements that do not fit the understanding the creators are trying to impose. The need to standardise is the need to reduce, to make the control of the data, and hence people, possible. Neil Postman states, 'we may say that in principle a bureaucracy is simply a coordinated series of techniques for reducing the amount of information that requires processing. Beniger notes, for example, that the invention of the standardized form –a staple of bureaucracy –allows for the 'destruction' of every nuance and detail of a situation. By requiring us to check boxes and fill in blanks, the standardised form admits only a limited range of formal, objective and impersonal information, which in some cases is precisely what is needed to solve a particular problem. Bureaucracy is, as Max Weber described it, an attempt to rationalize the flow of information, to make its use efficient to the highest degree by eliminating information that diverts attention from the problem at hand' (Postman 1993 p. 84). My work concentrates on subtly altering these forms through artistic mark–making and their layout structure. This meant experimenting through hand drawing elements of the form and manipulating the layout of both the text and boxes. A number of experiments were undertaken in recreating the form as a hand drawn object. *Barclayloan Application Form #1* (figure 1) was the first attempt at this. This series concentrated on altered loan application forms and tried to undermine the dehumanising process they represent even when advertising the process as the promise of 'a better life'.

Maya Lin was only 20 when she submitted the winning design for the *Vietnam Veterans Memorial*, simple polished black granite inscribed with the 57,661 names of the Americans who died in Vietnam. The monument is the most visited work of contemporary public art in the United States. After researching early memorials, she was struck by the World War I structures, many of which carry lists of the names of those killed. This is not reflected in later memorials where at a national level individuals are seldom acknowledged. She says, 'They captured emotionally what I felt memorials should be: honest about the reality of war, about the loss of life in war, and about remembering those who served and especially those who died' (Lin 2000). She was influenced by the lists of alumni killed in war carved into the walls at Yale and particularly by Sir Edwin Lutyens monument to the missing soldiers of the Battle of the Somme in Thiepval, France. The power of a name or the power to name is a key element in this and in Emin's work. Lin says, 'The use of names was a way to bring back everything someone could remember about a person. The strength in a name is something that has always made me wonder at the 'abstraction' of the design; the ability of a name to bring back every single memory you have of that person is far more realistic and specific and much more comprehensive than a still photograph, which captures a specific moment in time or a single event or a generalised image that may or may not be moving for all who have connections to that time' (Lin 2000 p. 4:10). The names of the dead are listed chronologically from 1955 to 1975. The moment of notification, when it happened is an emotionally compelling time. This work differs from Emin and Gordon's *List of Names* in that chronology, though not absent, is measured by the artist's life and not the subjects. However, where Lin and Emin parallel is in their response to the usual desire to cover up or not acknowledge that which is painful or unpleasant. Again we must look to the truth or, rather, self–regulated integrity. Memorials that have this integrity always appear to generate controversy. Rachel Whiteread's *Nameless Library* is a concrete block that shows shelves of books with their spines turned to the inside, the centre inaccessible by a permanently locked door. Whiteread states that it represents all the Jewish culture and learning which was lost forever in the Holocaust. Simon Wiesenthal, the veteran Nazi hunter responsible for the commission, says the image of the book has a special meaning for Jews. 'We are a people of books. We did not build our monuments out of stone and metal. Our monuments were books' (Masterman 2000). In Whitread's sculpture it is not the names of the dead but the names of the concentration camps in which 65,000 Austrian Jews were killed that are engraved around the base of the monument. This gives it an external meaning or at least one that is external to the subject. Emin's work is about herself; Lin's about the people who died but Whiteread's is about the perpetrators. It's a role of infamy.

Richter's work is perhaps the more difficult as it has fewer of the formal clues that exist within Emin and Lin. As mentioned earlier, there are aspects of the form of both Emin and Lin that fit into context, primarily the use of text for listing or cataloguing. Gerhard Richter's 1988 series *October 18, 1977* does not contain information in the same way the original photographs do now or did in their original context, i.e. representative data about the act. Do they contain information in any useful or definable form? Herman Hollerith, the founder of the Tabulating Machine Company that would later become IBM, got his idea for the punch card to tabulate census data from the use of train tickets punched in a specific pattern to record physical characteristics such as height, hair colour and clothing. He thought of these as a sort of 'punched photograph' (Black 2001 p. 33). The 'code' could then be read by a conductor, reducing fraud from passing the ticket to another person, in practice a simplified Bertillon system. These were a form of visually transformed representation, though in IBM's case in World War II, used to organise Germany's transportation, categorisation and extermination of the Jewish people and other 'undesirables'. Richter's work responds to the desire to identify and classify, specifically by the state that re–emerged as a result of the type of terrorism perpetrated by Baarder–Meinhof. 'For the first time since World War II, Germany collected massive quantities of information on its citizens –by 1979, the BKA's computer centre in Wiesbaden had files on 3,100 organizations and 4,700,000 individuals, along with 2,100,000 sets of prints, 6,000 writing samples and 1,900,000 photographs' (Storr 2001 p. 55) The 15 paintings in the 18 October 1977 series are all monochrome works. In one sense Richter is removing information, which leaves us with less a feeling of death crash voyeurism, as in Warhol's silkscreen prints, than 'exemplary suffering'. In this we may draw parallel's with Lin's Memorial. As the images are blurred we find we cannot supply the missing details or indeed correct the incomplete resolution (ibid pp.111–12). As Richter says, 'Life communicates itself to us through convention and through the parlour games and laws of social life. Photographs are ephemeral images of this communication –as are the pictures that I paint from photographs. Being painted, they no longer tell of a specific situation, and the representation becomes absurd. As a painting, it changes both its meaning and information content' (Obrist 1995)

The relationship between 'information' and 'art' should not be expected to be fixed or all–inclusive. Indeed. it may be that Information Art is a sort of oxymoron and its interest lies there. The most interesting area for me is in the interplay of the two different modes of interaction that these artists have with the world, the systematic or technical, exemplified by correctness, rules, symbols, etc., and the creative, embodying abstraction, richness of meaning and gestalt. There is need of a wider discussion about 'artistic truth', 'authenticity' and 'legitimacy' and how they fit within the arc of data, information, knowledge and wisdom.

> *The Information Man is someone who comes up to you and begins telling you stories and related facts about a particular subject in your life. He came up to me and said, 'Of all the books of yours that are out in the public, only 171 are placed face up with nothing covering them; 2026 are in vertical positions in libraries, and 2715 are under books in stacks. The most weight on a single book is sixty–eight pounds, and that is in the city of Cologne, Germany, in a bookstore. Fifty–eight have been lost; fourteen have been totally destroyed by water or fire; two–hundred sixteen books could be considered badly worn. Three hundred and nineteen books are in positions between forty and fifty degrees. Eighteen of the books have been deliberately thrown away or destroyed. Fifty–three books have never been opened, most of these being newly pur-chased and put aside momentarily. Of the approximately 5000 books of Ed Ruscha that have been purchased, only thirty–two have been used in a directly functional*

manner. Thirteen of these have been used as weights for paper or other small things, seven have been used as swatters to kill small insects such as flies and mosquitoes, two were used as a device to nudge open a door, six have been used to transport foods like peanuts to a coffee table, and four have been used to nudge wall pictures to their correct levels. Two hundred and twenty–one people have smelled pages of the books. Three of the books have been in continual motion since their purchase; all three of these are on a boat near Seattle, Washington.' Now wouldn't it be nice to know these things?

(Coleman 1972)

Bibliography

Black, E. (2001). *IBM and the Holocaust*. London: TimeWarner.

Button, V. and Esche, C. (2000). *Intelligence New British Art 2000*. London: Tate Gallery Publishing.

Coleman, A. (1972). 'My Books End up in the Trash', in Ruscha, E. and Schwaqrtz, A.(eds.). (2002). *Leave Any Information at the Signal: Writings, Interviews, Bits, Pages*. London: MIT Press.

Elsner, F. and Cardinal, R. (eds.). (1994). *The Cultures of Collecting*. London: Reaktion.

Lin, M. (2000a). *Boundaries*. New York: Simon and Schuster.

Lin, M. (2000b). *Making the Memorial*. Retrieved 18 December 2001 from www.nybooks.com/articles/13884

Masterman, S. (2000). *Remembering*. Retrieved 18 December 2001 from abcnews.go.com/sections/world/DailyNews/memorial.html

Obrist, H. (ed.). (1995). *Gerhard Richter: The Daily Practice of Painting: Writings and Interviews 1962–1993*. London: Thames and Hudson.

Postman, N. (1993). *Technopoly: The Surrender of Culture to Technology*. New York: Vintage Books.

Storr, R. (2001). *Gerhard Richter: October 18, 1977*. New York: Museum of Modern Art.

4.3 Metaphorical Vestiges on Info–Viz Trails

Donna J. Cox

Metaphor is not just a linguistic trick or Queen of Tropes; discourse on metaphor has engendered a paradigm shift in the way that we think about creative thinking. Over the last five decades theories of metaphor have rippled in some fields and roared through others. Researchers have generated more than 10,000 articles and books on metaphor and related studies. In particular, most of these writings have concentrated on linguistic metaphor and its connection to cognitive processes. Borrowing from the definition of George Lakoff and Mark Johnson (Lakoff and Johnson 1980), metaphor is more about thinking than about language. Metaphor involves the cognitive process of understanding one domain of information in terms of another domain. For example, 'man is a wolf' is a metaphor where a new understanding of man is provided by the cognitive mapping attributes or characteristics of the concept of 'wolf' onto the concept of man. The domain of information about man is understood in terms of the domain of information concerning wolf. To expand upon this basic idea, Lakoff and Johnson have delineated thousands of linguistic metaphors that we use to understand information. They defined 'conceptual metaphors' as those linguistic connections that have become so useful and common that they have now become understood as literal language. Such a metaphor as 'time is money' is integrated into the American culture to the point that we understand time in terms of money and conceptualise time as being 'spent', 'saved' or 'wasted.' Basic conceptual or conventional metaphors structure our everyday thinking. We interpret these metaphors literally as a part of speech. For example, 'argument is war' is basic and formulates how we verbalize argument: we defend, strategize, attack and defeat arguments. Conceptual metaphors shape the realities that we live by shaping our understanding of basic concepts that we attempt to verbalize.

A primary characteristic metaphor involves how novel they are. The range from literal or 'conventional' metaphors to the more novel and figurative metaphors is called the metaphor–content continuum. Verbal metaphors such as 'books are fresh fruit' are considered more figurative than 'his time was well spent'. Time being understood in terms of money has become embedded as 'conventional' language in American culture and, thus, is interpreted as literal language. The greatest bone of contention is what constitutes the 'conventional' end of this metaphor content–continuum. Most theorists agree that fresh, novel verbal metaphors can eventually evolve within culture and move from the novel end of this continuum to the conventional, as cultural accommodation reduces novelty to literality. Lakoff has primarily been concerned with the verbal conventional end of this range, while visual metaphor theorists primarily focus on the figurative or novel end. I will provide examples across the range.

One example of how visual metaphors can influence our understanding is through the development of information visualisation. Information visualisation is a relatively recent outgrowth of computer graphics and proposes to place abstract statistical data into a digital visual form that people can understand. Visualisation is the process of using computer–mediated technologies to display quantitative and qualitative information. More specifically, scientific and information visualisation is the process of transforming a system of mathematical and scientific models, statistics, assumptions, theories and other data into animated and interactive visuals that are rendered using two–and three–dimensional computer graphics. I use the term info–viz to specify this process of transforming scientific or information data and to distinguish info–viz from other visualisations. The resulting artefacts from info–viz will be referred to as 'visaphors', to distinguish them from other visual metaphors. Visaphors can be interactive within software environments or animated on a variety of display devices that range from stereo–CAVES to IMAX movies. Visaphors provide a rich resource for discussion in terms of metaphor and cognitive theories because visaphors bring to the foreground important issues that include literal interpretation, truth and perception; and they have a specific, self–evident intent by their creators to convey digital information in visual form.

Info–viz is a way of organising the incoming 'kaleidoscopic' flood of perceptions and concepts. 'Quantity' is a concept that we use daily and quantifying physical experience for cognitive ends is important to organizing our world. We want to understand what proportion of people die at our age, how many calories affect our bodies, how much automobile fuel will cost and what fills our cup. Understanding information in visual form is motivated beyond academic inquiry. The graphical display of quantitative information is important to culture and the individual and to ignore its position in the history of metaphor and cognitive theories is a waste of natural resources. Information in quantity form permeates our society. Thus for the purposes of this paper, I will refer to information visualisation as info–viz and the resulting visual metaphors as visaphors.

While there has been extensive research on verbal metaphors, few have focused on the visual counterparts though visual metaphors subtly shape our realities as linguistic metaphors have been demonstrated. One researcher, Charles Forceville (1996), analysed pictorial metaphors in advertising. Forceville focuses on figurative, static images and deals with issues of mass communication. Advertising has the self–evident goals of communicating a message to sell a product. Info–visualisations often have the self–evident goal of conveying information and promoting ideas. This comparison reveals the similarities and extensions to the linguistic non–arbitrary metaphor 'mapping function': one domain of information maps onto another domain to provide a new relationship of meaning.

In a recent project, IntelliBadge, we tracked volunteer participants at an academic conference. We measured and tracked the data of people's profiles through a three–dimensional space of the Baltimore Convention Center during a Supercomputing Conference. People carried Radio Frequency (RF) tags during the conference where the Convention Center was instrumented with technology that measured their physical locations. I contrast two different visaphors showing the same real–time tracking and categorical data. We used a dynamic bar chart and map as a more 'literal' metaphor. Compare this visualisation to the 'How Does your Conference Grow' metaphor. The domain of information being visualised for both visaphors is the same but not present: a live database, radio frequency signals, people moving throughout the building, their professional aggregated interests, the rooms being tracked, etc. Thus two visual metaphors represent the entire span from the literal bar chart to the figurative flowers. Both represent the same data, but are on opposite ends of the metaphor content–continuum.

Measuring metaphors and visual devices for charts and graphs all have had novel beginnings. Many of these devices have evolved into sets of consistent visual metaphors. Maps themselves are good examples of how conventional visual metaphors have developed into coherent and consistent systems of visual metaphors. Here I contend that modern maps and many visaphors are visually 'read' through visual *conventional* metaphors (in the Lakoffian sense) and that their novel origins have been lost over time (owing to familiarity). The histories of map–making before the sixties show a cultural bias of western European thinking (Bagrow and Skelton 1964). The projection systems, lines on maps and scales can be traced back to a disciple of Aristotle and Greek mathematics. We literally interpret such conventions today, though their origins reveal one set of visual metaphors to represent maps. Indigenous peoples navigated land and sea with competency and accuracy. Indigenous peoples used a different variety of materials and visual idioms. Early cartographer's criticisms 'that these savages couldn't draw in perspective' are unfounded. For example, the Marshall Islanders designed intricate patterns from palm fibre and shells to represent wave–crests, navigation sites and mariner's direction. These accurate and useful maps provided alternate visual metaphors for navigating land and water. Likewise, the Aztec and Mayan maps were accurate though employed different projections and models. Cortez conquered the great Aztec civilization using their indigenous maps painted on cloth. Afterwards Catholic friars systematically destroyed these early meso–American maps and replaced them with Spanish western European maps. Map–making has primarily been dominated by western European thinking. We can locate a variety of visual methods that enabled different groups of people to navigate their terrain, find locations in the land they travelled and explored, and mark their territories with different measuring devices. In *Guns, Germs, and Steel* Diamond argues that indigenous peoples are as intelligent as white people with lots of technology though there are differences in the way of describing the world and the motivations of the cultures. Diamond points to factors other than superior mathematics or technologies that provided advantages to the Mediterranean cradle of civilisation. Today's methods for spatially quantifying information have roots in civilisations that are not smarter but more dominant in terms of the source domain. As we navigate the highways of our life, it is possible that finding our location in space would involve a very different set of symbols, projections and visual idioms had Aboriginal cultures colonised the world instead of the Europeans.

Contemporary astronomy and astrophysical maps and models of the universe fit into the evolution of cartography, scientific thinking, and theoretical studies. The weight of science and technology behind the visaphors affects the audience's interpretation. The cultural context is often over television or museums. Little attention is given to the assumptions behind the concept network of beliefs, the inadequacies of technologies, perspective projection, incomplete scientific models, errors and computer graphics devices. These visaphors are positioned within the cultural, militaristic and political domains from which they evolve. Major cultural assumptions underlie the process of info–visualization. Some have argued that scientists and mathematicians work to maintain extremely coherent and consistent sets of metaphors. When we take these metaphoric relationships too seriously we undermine our creative thinking and our realities become calcified within consistent metaphoric myths. Through accommodation and cultural contexts, our realities are being created. Visaphors such as astrophysical or earth maps cannot be arbitrary; they have to work to a physical degree or we will not use them; however, alternative idioms have been abandoned for consistency and are culturally contingent. I contend that the process of info–visualisation is 'metaphorical in nature'. The 'metaphoric action' that I claim is not ritual; it involves mental imaging, cognition and the making tangible through technology the understanding of one domain of knowledge in terms of another. Info–visualisation is a metaphoric 'interaction' (not in the sense of ritual) and visaphors can range from literal to novel. There are visual counterparts to conceptual linguistic metaphors that are interpreted as literal and formulate our realities. Through accommodation, we are creating reality using validated scientific models. There are other ways to design visaphors but the mappings have to work. I challenge the info–visualisation community to employ artists and expand their literal devices.

Bibliography
Bagrow, L. and Skelton, R.A. (1964). *History of Cartography*. Cambridge, MA: Harvard University Press.

Forceville, C. (1996). *Pictorial Metaphor in Advertising*. London: Routledge.

Lakoff, G. and Johnson, M. (1980). *Metaphors We Live By*. Chicago: University of Chicago Press.

Diamond, J. (1999). *Guns, Germs, and Steel: the Fates of Human Societies*. New York: W.W. Norton

4.4 Interstellar Messaging, Xenolinguistics and Consciousness: LiveGlide Meets the SETI Enterprise

Diana Reed Slattery and Charles René Mathis

Introduction

SETI asks: if we do receive, distinguish –and decode –a message from the stars, how –and what –should we reply? And further, should we be sending our own messages out in hopes of reaching an ETI (Extra–Terrestrial Intelligence)? If so, what do we wish to communicate? And with whom do we imagine we are communicating? The topic of consciousness permeates the SETI discourse in the assumptions underlying our answers to those questions. Imagining intelligence some significant degree different from, more 'advanced' or 'evolved' than our own, poses unique problems. Allowing the possibility of some significantly higher degree of intelligence (perhaps evidenced by superior technology or an ability to get along with each other), we try to imagine what such a mind –and states of that mind –might be like and might be after (goals, intentions, concerns). From there, one might wonder whether the ETI's forms of symbolic processing (language) might also have developed in some direction as far beyond human 'natural language' as we see natural language as more cognitively potent than, say, animal calls.

Epistemology

Rather than focus on the rich offerings of ideas of how we might communicate with ET, which can be found online in the abstracts of the participants,2 this paper examines the tacit epistemological assumptions within which this workshop, and SETI as a whole, seems to operate. SETI is firm on positioning itself as a scientific enterprise and distances itself from any and all personal reports of alien contact, abduction, Raelian cults and UFO sightings, using only one main data collection strategy: listening to the skies for signals in the electromagnetic spectrum (radio or optical) that might be distinguishable from background noise and identifiable as the communication of an ETI. Frank Drake, SETI pioneer and granddaddy of interstellar messaging, assumes this position when he states, 'Personally, I find nothing more tantalizing than the thought that radio messages from alien civilizations in space are passing through our offices and homes, right now, like a whisper we can't hear. In fact, we have the technology to detect such signals today, if only we know where to point our radio telescopes, and the right frequency for listening' (Drake and Sobel 1992, p. xi). Drake's intentions for his book *Is Anyone Out There?* are clearly given: 'I want to nurture people's yearning curiosity about the beings who will no doubt contact us. At the same time, I want to quell the misleading myths about extraterrestrials, from the mistaken belief that they have visited us in the past, to the terrifying idea that they will wrest the future from our hands. I want to show that we need not be afraid of interstellar contact, for unlike the primitive civilizations on Earth that were overpowered by more advanced technological societies, we cannot be exploited or enslaved. The extraterrestrials aren't going to come and eat us; they are too far away to pose a threat. Even back–and–forth conversation with them is highly unlikely, since radio signals, travelling at the speed of light, take *years* to reach the nearest stars, and many *millennia* to get to the farthest ones, where advanced civilizations may reside' (ibid p. xiii). Drake's certainty about these matters –especially our safety –depends on a method of investigation coupled with a description of space–time that keeps ET at a safe distance, in a vast unfolded universe with strict speed limits. Einstein's prison is the last word in our description of the structure of the physical universe, one that any ET, even those in 'advanced civilizations', cannot escape. There are practical (political) considerations for sticking to the EMS bands for the search, such as funding. SETI's $100 million from NASA in the 1990s depended on a hard (those on the other side say hegemonic) science approach. Even one's tenure as a Harvard professor can be challenged, as John Mack found out when he took alien abduction reports as a subject for scientific inquiry. As Jodi Dean, academic and author of *Aliens in America*, says, 'Perhaps because he is already credentialed through his position at Harvard, John Mack views the abduction phenomena as striking at the heart of the Western scientific paradigm.' Citing a conversation with his longtime friend Thomas Kuhn, Mack explains that Kuhn helped him to understand that our current scientific worldview 'had come to assume the rigidity of a theology, and that this belief system was held in place by the structures, categories, and polarities of language, such as real/unreal, exists/does not exist, objective/subjective, happened/did not happen' (Dean 1998 p. 57). Whatever one's paradigm or 'politics of ontology' (Mack's fortunate phrase), strict adherence to one method of listening for signs of life in the rest of the universe and one select group of qualified listeners (SETI scientists) has, despite a $100,000,000 budget, led to exactly no data. This condition of 'no data' is established internally by the absence of suitable incoming signals. 'No data' is reinforced externally by the strict guards at the perimeters of the discussion that ensure no discussion –*no mention even*, if the Paris workshop is any indication –of what is going on in the wannabee field of ufology, to say nothing of 'the popular mind', will be made. The standard for reliable witness is simple: if you say you experienced something, you didn't. By definition.

Rather than discourage the SETI researchers, this lack of data opens, once one has accepted that this is an elite and gated community, an epistemological (un)commons of great freedom. In the absence of any fact, any report, any evidence or suspected evidence, one is free to speculate, amazingly free, free as –say –a science fiction writer. Science fiction tends to follow a formula of 'what if' –taking a scientific fact, or theory, or technological development, and extrapolating a scenario, often a whole self–consistent world, based on a logical unfolding of ideas but generally anchored in some notion (no longer necessarily from the physical sciences alone) currently accepted as real, or possible, or at least discussable. The notions offered at the Paris workshop were similarly grounded in some area of current knowledge –interspecies communication, shamanic journeying and cosmologies, anthropology, evolutionary algorithms, music, Kabala, logic and theology, to name a few. Additionally, the lack of data gives SETI the rhetorical flavour of a messianic cult, where the faith that we are on the verge of an arrival thrives undaunted (Drake and Sobel 1992 p. xii),[3] not only in spite of the absence of data but where the deferral of consummation is essential to the survival of the devout wish. Not to mention that the institutional survival of a group organized around a *search for* would be in danger should the search successfully conclude.

Glide as xenolinguistics

My offering to the workshop was Glide –a language originated in the novel *The Maze Game* (Slattery 2003). Glide is a discrete system of signs or glyphs that can link and morph dynamically. Each glyph contains a set of meanings and those meaning merge as glyphs transform into or combine with other glyphs.

Current research includes two new applications that move the writing space from planar inscription into the 3rd spatial dimension.

Using the SETI *gedanken* of communicating with ETIs as a framework, I raised questions about language and consciousness. Is human language itself evolving beyond the structures of spoken or written 'natural language'? If so, how might our cognitive processes change, especially if the fundamental structure of a new kind of language were not primarily propositional but, say, metaphorical? If the symbols were expressed in multiple dimensions and through multisensory modes of representation, what kinds of complex and nuanced communications could be simultaneously apprehended? Can we extrapolate the necessary rewiring in neural circuits in the infrastructure of the brain/mind to support a different symbolic system from what we know about neural plasticity, or the capability of the brain/mind to produce radically altered experiences when tweaked by meditation, a different mix of neuro–transmitters (Munn 1973) or beamed with ELF/VLF frequencies?[4] Glide, originally a gestural language, allows expression of emotion and attitude through body language, the non–verbal component of human communication by which we assess each other's intention, including placing 'the other' on various scales regarding their intentions: Dangerous or safe? Altruistic or selfish? I offered the idea that use of a model like Glide to construct an interstellar message may, at base, communicate something fundamental about ourselves as a species with a rich and diverse linguistic capability that extends not only through the huge variety of natural languages but into visual, gestural and 'body' language as well. In displaying this capability, we could communicate something fundamental about what it means to be human in the universe, the prose *and* the passion, our knowledge as well as the desire we hold for communicating and for making connections across the interstellar distances.

Xenophobes and xenophiles

At the end of the workshop a heated discussion arose as to whether, after so many years of silent listening, we should we communicate 'pre–emptively' with ETI. The issues that split the group came down to 1) Is it safe? 2) Who is authorized to communicate for the whole human race? 3) What message should be sent? The initial premise of how to encode altruism was often set aside in this part of the discussion as suggestions of how to disguise ourselves, prevaricate and hide our position and intentions from a potentially hostile ETI were considered. One possible interpretation of the fact that we have received zilch on the EM spectrum that could be considered a signal could be because any ET would have learned from experience to hide its electromagnetic culturally generated radiation and not make itself a blatant target. When push comes to shove, ETI no longer could be depended on to be at a safe distance, not to have advanced beyond our level of technology –or our inability to imagine the possibility of escape from Einstein's prison. The imaginations of the xenophobes moved easily into the well–worn runnels of paranoia and defensive strategies. Next step: don't communicate at all. Hunker down and hold onto the hope nobody's home in the universe but us. Regarding the matter of who was authorized to speak for 'us', the artists tended to assume they had the right to communicate freely about anything to anybody in any format, whereas others felt that government(s) were in charge of communications, although the prospects of reaching global agreement on such a matter seemed more remote than achieving faster than light travel. To trust or not to trust? Of course, if hypotheses such as Francis Crick's directed panspermia (Crick 1981), Fred Hoyle's comet seeding (Hoyle and Wickramsinghe 1979) or Svante Arrhenius's spores travelling through space (Arrhenius 1908) are on the table, then, as Pogo said, 'We have met the enemy and he is us.' Our message, in such a worldview, with a dawning awareness of our long exile, might be reframed as 'ET Phone Home'.5 The Paris SETI workshops could become sites for an emerging awareness that we are not only not alone but not at home, that Gaia is a stepmother and that, like many adopted children, we can't help wanting to know who our biologically real parents are.

A Final speculation

SETI is one of the great *gedanken* of our dreaming species. As scientists, artists or science fiction writers, we speculate on a level playing field. True, there are reports out on the fringe that the Space Brothers have already called in. So far such reports have only been investigated as psychological or sociological anomalous phenomena:

not admissible evidence by those whose cultural role it is to issue certificates of authenticity to experience. Therefore, in the absence of any confusion or restrictions caused by validated facts, each of us is equally authoritative as to the nature, capabilities, the agenda, even, of an ETI. With that disclaimer, this professional speculator will share a final imagination about the use of Glide as a means of reaching the ETIs we hope and/or fear might hear us.

Glide began as a narrative element in a work of speculative fiction, *The Maze Game*, before it jumped out of the story world to become an object of linguistic inquiry, an artefact to be examined intellectually and experimented with in a series of software applications (www.academy.rpi.edu/glide). After considerable play and examination of the structures and logic of Glide, the author got the notion of using Glide as a means to communicate with ETIs and connected with the SETI effort. *Somewhere in the universe, the Glide Council smiled.* They had felt frustrated in many previous attempts to get through to the human species using more direct —mind to conscious mind —channels of communication which seemed to either frighten the recipients (who shouted *abduction*!), turned them into gurus channelling the Zeta Reticulins or had them committed to psychiatric care. The Glides then ramped down their communications to arrive on a more acceptable channel —out of the great dark grab–bag humans call the subconscious. Humans, it was noted, accept dreams, stories, even scientific realizations from this dark source and seem content to examine such fully formed creations without worrying about who put them there or how they came into consciousness exactly when they did. Through this convenient backdoor into the human psyche, the Glides then downloaded their living language, embedded in a story, into a receptive human mind. The viral meme of the Glide signs took over the story itself, fed on the rich brew of notions in the host, rearranged the story to include an explanation of itself and its origin and achieved escape velocity into a status independent of the story. The Glide language, being (in the eyes of the Glides, at least) a tad more advanced than human languages, has the property of being able to teach itself, possessing a degree of livingness (ability to move, gesture, transform). The Glides counted on being able to move and transform the human recipients with their language, as language has always been the technology by which the humans not only communicate but shape their conscious experience. So, the Glides reasoned, if we can engage the curiosity of the humans and entice them to pick the Glide signs up and play with them, perhaps it will occur to them to communicate in our direction, using our very own language. Then if we confirm the transmission using the same means, perhaps they'd know they'd really connected after all.

Bibliography

Arrhenius, S. (1908). *Worlds in the Making: The Evolution of the Universe*. London: Harper and Brothers.

Crick, F. (1981). *Life Itself*. New York: Simon and Schuster.

Dean, J. (1998). *Aliens in America: Conspiracy Cultures from Outerspace to Cyberspace*. Ithaca: Cornell University Press.

Drake, F. and Sobel, D. (1992). *Is Anybody Out There?: The Scientific Search for Extraterrestrial Intelligence*. New York: Dell Publishing.

Hoyle, F. and Wickramsinghe, N.C. (1979). *Lifecloud: The Origin of Life in the Universe*. New York: HarperCollins Publishers.

Munn, H. (1973). 'The Mushrooms of Language', in Harner, M. *Hallucinogens and Shamanism*. Oxford: Oxford University Press.

http://publish.seti.org/art_science/2003/

Slattery, D. (2003). *The Maze Game*. Kingston: Deep Listening Publications.

Notes

1. Charles Ren_ Mathis provided the programming for LiveGlide, involving complex real–time computations of morphing 3D geometries.

2. http://publish.seti.org/art_science/2003/

3. 'I am telling my story because I see a pressing need to prepare thinking adults for the outcome of the present search activity –the immanent detection of signals from an extraterrestrial civilization. This discovery, which I fully expect to witness before the year 2000, will profoundly change the world' (Frank Drake 1992).

4. Refer to the work of neuroscientist Michael Persinger, at www.laurentian.ca/neurosci/persinger.html. Accessed 14 June 2003.

5. For immediate practical assistance one can visit Aliens On Earth.com: Resources for those who are stranded here www.ufomind.com

4.5 Art and HCI: A Creative Collaboration

Ernest Edmonds, Linda Candy, Mark Fell and Alastair Weakley

Introduction

The Creativity and Cognition Research Studios (C&CRS) were established for the purposes of developing new art and technology projects and to conduct research into the creative process. It is the result of a collaborative venture between the Department of Computer Science and the School of Art and Design at Loughborough University. In the COSTART Project 1998–2003 an innovative approach to the study of technology–based art founded on practice–led research methods has been developed. The approach is based upon artist–in–residency studies, gathering and analysing qualitative data and disseminating new knowledge on the basis of the evidence (Candy and Edmonds 2002, 2002b, 2002c; Licht 2002).

Research issues that are under investigation at C&CRS include: the impact of the technology on creative practice, the implications of such practice for technological requirements and the environments in which new developments can take place. The work centres on the practice of art making.

The involvement of the artists in the electronic media is as much concerned with developing and defining those media as with employing them in art making. Whilst art and design oriented application programs are often used, the art practice is normally dependant upon writing computer programs, often interfaces between the various devices needed to facilitate interaction. The artists at the leading edge of technology–based work are rarely confined to using single software applications that can be bought on the high street. Instead, they are most often seen to be extending the media and exploring the means of developing new technological capabilities. As is often the case in innovative art making, artists are deeply involved in inventing and defining the media that they use. If this requires new knowledge and skills, they acquire it either by learning it for themselves or through their collaborators. For this reason, collaborations with technical experts who can construct and extend the technology are becoming a vital element of the work. In this respect, a significant development is the invention of computational representations of the conceptual and behavioural concepts that underpin much of the art being developed.

In this paper we report on the particular innovations associated with one of the residencies –that of Yasunao Tone, an artist who was awarded the 2002 Ars Electronica Golden Nica prize for Digital Music (Licht 2002). The residency involved three significant interacting roles: the artist, the technologist and the researcher. This three–pronged approach is fundamental to the research and has been significant in enabling and fulfilling the kind of innovation described here.

A residency study

Yasunao Tone is often associated with the Fluxus group that began in the 1960s (Friedman 1998). Throughout the 1960s and 1970s Fluxus evolved into an international art movement, notorious for its refusal to accept traditional ideas about art, culture and authorship. Through a series of events, happenings, performances and installations Fluxus became one of the most radical movements in contemporary art of its day.

As the artist put it: 'I have pursued totally new relationship between text and sound aside from traditional lyric–melody etc. and more trendy textual music as well.' *Molecular Music* (1982–5) is an earliest experiment as such. The piece is based on the poems written in Chinese characters including three poems from the Tong Dynasty as well as the 8th century Japanese poem. The Chinese characters of the poetic text are grammatologically studied first and, then, appropriate images are chosen from found photos. Then, they were filmed as if an animation movie was taken. Rhythmic structure of the spoken text is transferred to the structure of the film. The piece employs the sound–generating system including light sensors attached on the film screen and oscillators connected to light sensors, so that the film projected on the screen creates varying sounds in accordance with the specific arrangement of the sensors and the changing brightness of the projected images.

Another case in point is *Trio for a Flute player* (1985), which consists of three components. These components are based on a single source, poems from the 8th century Japanese anthology, the Man'yo–shu. The curvy line of calligraphy of the poem, overlaid by musical staff, does not correspond to pitches or any tonalities but with the player's finger placements. (Note that a flute player uses nine fingers, coinciding with the number of lines and spaces of the staff –five lines and four spaces). Fingering, with its movement and pressure, triggers an electronic sound, varying in pitch and intensity, which is generated by an oscillator with a capacitor. The poems are read through the flute mouthpiece.

Yasunao Tone's process for working with the conversion of calligraphic drawings into sound is very time–consuming. involving as it does changing first Chinese characters into images and then the images into sound. Such timescales restrict the use of real–time interaction with his work. The aim for the COSTART residency project was to explore the possibilities of sound representation in real–time. The intention was for the development of a programme for performance that could focus on the interaction between image and sound. The Soft–board made it possible to do live performance in which, instead of having to transform the text into images with pictogram–like Chinese characters, the artist draws calligraphy on the board and transforms any text into sound.

The concern throughout the residency was how working with an artist of this nature influenced the process of making the work and how a rejection of prescribed methods, tools and structures changed the way in which software systems were developed. This activity was further complicated because the apparent divide between the work and the means of making the work. So there was no point at which we were making simple tools or technical solutions that would later be used by the artist to produce the work itself. The work was both the system and the process as much the outcome of these activities in a performance.

In the pre–residency discussions for Yasunao Tone's residency, it had seemed that it would be interesting to make data about his drawing gestures available computationally. The idea was that he would draw on the Soft–board[1] in our studios. This device looks like an ordinary whiteboard but has the additional capability to detect the position of a pen on its surface. This information is output via a serial cable and can be used to keep a dynamic record of developments in a meeting, for example. This was an ideal way to explore Yasunao's use of calligraphy and how it can be transformed into sound.

When it was suggested to Yasunao that a variation on granular synthesis to generate sounds could be used, he said that his idea was more like 'macaroni synthesis, because it already had a form'. Tone explained to the COSTART observers in the final interview of the residency what he meant by that:

YT   Oh!... I have a piece called 'Molecular Music' which... the title itself is 'My Dissatisfaction with Computer Music'. Computer Music is a kind of... started from atom. Yes... so I start from molecules not from atoms. And this time I told Mark, 'This is a macaroni'....it's developed through granular synthesis. Granular synthesis is... he obtains certain small tiny bits of a wave form and in the building, so doing you have to make a pitch how you make yours, you have to do the same thing as I know, take academic computer music. So I don't like that. So instead of grain I use macaroni.
LC   Well, there's some cooking required after that.
YT   Yes cooking.
LC   Is it cooked macaroni first or... the basic ingredients?
YT   Yes, basic ingredients.
LC   The performance is the cooking?
YT   Right.

When the team tried to imagine a system that transforms the movement of a pen into a sound they almost inevitably started to think in a certain way –to define a complex system of conditions, choices, relationships, behaviours, where a given input is mapped to a particular parameter and a certain kind of intervention has a certain kind of result. This kind of method seems true of most explorations of correspondence between image and sound through interaction.

According to this approach a perfect system might be one that is flexible enough to generate an almost endless series of tonal variations yet be controllable enough so that each one makes musical sense. In making such a system we primarily concerned ourselves with two types of question, firstly about how data is mapped to sound and secondly how we might operate the system, how we know what it will do under certain conditions and how we become skilled in its use. The ultimate assumption is that the system is a tool or instrument of expression and that this knowledge will enable the artist to control the system, so that he or she can articulate and communicate something about his or her feelings or beliefs. There are works that employ the opposite approach, ones that make the relationship between input and output problematic by creating systems that range from the seemingly random to ones whereby the users grasp some sense of what they are doing but are unable to understand fully.

It was known in advance that Yasunao's work embodied the polar opposite of these issues. Accepted knowledge gained in addressing such issues would be called into question (or even totally ignored). Such work not only challenges ideas about art, technology and creativity but in doing so presents new perspectives on sound and image correspondence, generative systems and human computer interaction in the widest sense. However, these kinds of issues are not of central concern here, notwithstanding their interest.

A new instrument for interaction

The interaction device made during the residency uses the Soft–board upon which the artist draws a series of strokes. The Soft–board sends information about pen colour and pen position to software that is used to synthesize sound. Projected onto the Soft–board is a sequence of video images selected by the artist. As the artist draws, the video image advances frame by frame. Data is taken from the x–y coordinates, the speed of movements and mapped to synthesis parameters. For the duration of the residency the sound was relayed using a pair of Tannoy active reveal monitors playing at high volume.

The system was developed and implemented in Max/MSP, which is a graphical programming environment made by Cycling 74. At the time of making the work the third author was an Alpha Tester for the Jitter objects for Max. These enable the integration of real–time video and 3D graphics into the Max environment. The Jitter objects were developed by Joshua Kit Clayton at Cycling 74.[2] It was decided to make an effective interpreter within Max/MSP. This object receives the stream of data from the Soft–board and outputs information about the colour of pen being used, together with its position on the surface of the board and information about when the current drawing stroke began.

The Soft–board is connected to the computer using a serial port. It sends a collection of items of raw data as described here:

- information about pen colour: black, blue, red, green or eraser

- the x–y coordinates of the pen position when it is placed on the drawing surface

- after an initial pen down, the delta position is sent at a regular interval until the pen is lifted off the drawing surface

- the time that the pen comes into contact with the drawing surface

From this raw input we were able to calculate the absolute position of the pen while it is in contact with the surface and also the speed or velocity of movement.

A video file was also loaded into the system. Jitter was used to integrate video playback. At each pen down event the video was advanced by one frame. The position of the pen is used to select a specific pixel of the frame. The brightness of the pixel is read and used in the synthesis algorithm.

When the pen is placed on the surface, the volume is switched on and when the pen is lifted off the volume is silenced. Thus sound is only heard when the pen is on the surface. When the pen is placed at the outer extremes of the board along the x axis (to the far left and far right) only the fundamental frequency is present. As the pen moves closer to the centre the harmonic content is changed: the first harmonic is introduced, then the second and so forth. The speed of movement in the x axis (horizontal) is used to determine the volume of sound. Slow movements in this direction create quiet sounds and fast movements create harsher sound. At the very centre of the Soft–board drawing area a small square is defined. When this area is drawn in it produces white noise.

Conclusions

In the Yasunao Tone residency a highly challenging view of human–computer interaction was taken. It led to the development of a quite novel interaction device that formed the core base of a new creative performance artwork by Tone.

The nature of the study undertaken, in which the roles of artist, technologist and researcher were allocated equal place, was highly effective in stimulating innovation. As well as enabling valuable research into the creative use of technology, the approach can be used to stimulate the creative development of technology.

Bibliography

Baudrillard, J. (1997). *Fragments Cool Memories III, 1991–95.* London and New York: Versa.

Candy, L. and Edmonds, E. A. (2002a). 'Modeling Co–Creativity in Art and Technology'. *Proceedings of Creativity and Cognition.* New York: ACM Press.

Candy, L. and Edmonds, E. A. (2002b). *Explorations in Art and Technology.* London: Springer Verlag.

Candy, L. and Edmonds, E.A. (2002c). 'Interaction in Art and Technology'. *Crossings: Electronic Journal of Art and Technology,* vol. 2, no.1 http://crossings.tcd.ie

COSTART, project: creative.lboro.ac.uk/costart

Creativity and Cognition Research Studios (C&CRS) www.creativityandcognition.com

Edmonds E. A. and Candy L. (2002). 'Creativity, Art Practice and Knowledge'. *CACM,* vol. 45, no. 10.

Friedman, K. (1998), *The Fluxus Reader.* Academy Editions. Chichester: John Wiley and Sons.

Kostelanetz. R. (1968). *The Theatre of Mixed Means.* New York: The Dial Press.

Licht, A. (2002). 'Random Tone Bursts'. *The Wire,* issue 223, September, pp. 30–3.

Max/MSP www.cycling74.com

Notes

1. Soft–board, now marketed as an *LT Series Interactive Whiteboard* www.polyvision.com/.
2. Max/MSP www.cycling74.com.

5. The Art

5.1 The Idea Becomes a Machine: AI and A–Life in Early British Computer Arts

Paul Brown

'The idea becomes a machine that makes the art' –Sol LeWitt.

(LeWitt 1967)

In 1968 I was one of a generation of young artists who visited the ICA (Institute of Contemporary Art) at their then–new premises in London's Mall to see the Cybernetic Serendipity show (MacGregor 2002). Curated by Jasia Reichardt Cybernetic Serendipity, like other exhibitions in the ferment of the 1960s, it challenged many long–held attitudes to the visual arts and their place in culture and society. In particular works by scientists were shown alongside those of professional artists and Reichardt did not differentiate, at least on the level of the exhibited artefact, between the 'two cultures'.

Like many of my contemporaries I was enthralled by the show and, after a period working with video and analogue electronic systems, I have since 1974 worked almost exclusively with computers and digital systems. Even younger artists, like Ken Rinaldo, credit the show for inspiring their interests in what Kay termed 'the computational metamedium' (Kay 1984). Rinaldo saw the show as a child and only later, when he discovered the catalogue in a second–hand art bookstore, recognised what he had seen and the influence it had had on his development as an artist.

One of the works shown at Cybernetic Serendipity was Edward Ihnatowicz's 'Sound Activated Mobile' or SAM. It consisted of four parabolic reflectors formed like the petals of a large flower on an articulating 'stem' or neck. Microphones placed at the foci of the reflectors enabled SAM to accurately detect the location of sounds and to track them as they moved around the exhibition. The visitors were left with an uncanny sensation of being 'watched' as they walked around. Alex Zivanovic's web site dedicated to Ihnatowicz's work (Zivanovic 2003) contains a video of SAM in action.

Later Ihnatowicz worked on the ambitious and high budget 'Senster' for the Phillip's Evoluon museum in Eindhoven in the Netherlands. Kees Stravers maintains a website about the Evoluon (Stravers 2003). The Senster was a 16–foot (4 m) articulating robot arm that responded to both sound and movement and was installed in the Evoluon's main entrance gallery. Originally programmed to respond to loud noises, it provoked a cacophony whenever school groups visited. This prompted the museum authorities to ask Ihnatowicz to reprogram it to interact with 'gentle' movements and sounds. It was installed in 1970 and decommissioned in 1974. A film of the Senster is also available from Zivanovic's web site (Zivanovic 2003).

Ihnatowicz's final robot piece was 'The Bandit', part-financed by the Computer Arts Society and exhibited at their show 'Interact' at the 1974 Edinburgh Festival. It was based on the familiar 'One-Armed Bandit' gambling machine. Visitors interacted with the lever and the system was able to make pretty accurate analyses of their gender and temperament.

During the period he worked on the robotic pieces Ihnatowicz was a researcher in the Mechanical Engineering Department at University College, London, which also hosts the Slade School of Art. In 1972 the late Malcolm Hughes, who was a member of the System Art group and head of Postgraduate Studies at the Slade, established their Experimental and Computing Department. Ihnatowicz was a frequent visitor throughout the 1970s and often engaged in informal discussions with staff and students on topics of interest. I remember one such discussion about Artificial Intelligence (AI). Edward referred to the work of Piaget on infant learning and the importance of the tactile stage that precedes and is an essential prerequisite for later visual and metric learning. He stated his opinion that if machines were ever to become intelligent they could only do so by interacting with their environment.

In retrospect it's possible to perceive that Ihnatowicz was an early proponent of embodiment in both the arts and AI and it's clear that he was also a pioneer of the discipline now known as Artificial Life (A-Life). Contemporary roboticists and AI specialists working in the now popular bottom-up methodologies (like, for example, evolutionary robotics) are often astounded to learn of Ihnatowicz's work, particular when they are told its early date. Ihnatowicz died in 1986.

Another regular visitor to the Slade's Experimental and Computing Department was Harold Cohen. Cohen was a well-established artist who had represented Britain with his brother Bernard at the 1966 Venice Biennale. In 1969 he began working at the University of California at San Diego (UCSD) where he became interested in computers and programming. From 1971 he was involved in the AI Lab at Stanford University where Edward Feigenbaum was developing 'expert systems'.

Expert systems get around a major problem in classical, top-down, disembodied AI research. The problem is context. The human mind has an amazing facility to quickly apply a multitude of contextual information to the cognition of ambiguities common in speech and other forms of inter–human communication. Even high–speed modern computers with their linear processing structures can't compete. Feigenbaum was one of a number of researchers who in the late 60s and early 70s suggested that this could be overcome by limiting the area of intelligence to small, well–defined knowledge bases where ambiguities could be reduced and the contextual cross–referencing applied.

Researchers at the Stamford lab developed many valuable expert systems like Mycin that was used to diagnose infectious diseases and prescribe antimicrobial therapy (Buchanan 1984). As a guest scholar and artist–in–residence from 1971 to 1973 Cohen began to develop an expert system he called Aaron. He continues to work on it and jokes that it's the oldest piece of software in continuous development (Kurzweil 2001). Aaron is a 'classical' top–down AI package. It contains an internal database and set of rules that enable it to produce sophisticated drawings. Although Cohen is interested in investigating issues to do with cognition and drawing, his major achievement has been the externalisation and codification of his own drawing abilities. Aaron produces 100 per cent genuine 'original' Cohen drawings without the need for the 'human' artist's intervention.

Ihnatowicz and Cohen represent the first 'great masters' of the computer–based arts and it's interesting that they also epitomised the two main approaches to Artificial Intelligence. Cohen's work builds upon the classical methods of 'top down' internal data representation and analysis. Ihnatowicz is an early pioneer of the now popular methods of 'bottom up' learning systems –an aspect of what's since become known as 'Artificial Life'.

The Slade's Experimental and Computing Department was strongly influenced by the European Systems Art movement. Many students were pursuing procedural methods for generating art. The lecturer in charge of the computer studio was Chris Briscoe who was interested in another important influence derived from contemporary scientific investigation –unpredictable, non–linear deterministic systems or what would later become known as 'Chaos Theory'. His drawings and sound pieces showed the evolving relationships between interdependent graphic and audio entities. Later in the 1970s a visitor from the USA, the Polish mathematician Andre Lissowski, introduced us to the work of Benoit Mandelbrot and his concept of fractals.

Another important visitor was Ernest Edmonds, an artist who had been using computers since 1968. As a professor of computer science, Edmonds founded the UK's main research initiative into human computer interface (HCI) first at Leicester and then at Loughborough. His work in the arts, as with his work in HCI, is concerned with embedded intelligence. His early paper on artworks that learn published with Cornock (Cornock and Edmonds 1973) was recently revisited (Edmonds 2003). Several of the computer studio students went on to pursue PhD studies under Edmond's mentorship and became some of the first visual arts students to achieve this award in the UK.

Major computational influences in the studio were Cellular Automata (CAs). These had been popularised by Martin Gardiner's piece about John Horton Conway's 'Game of Life' in *Scientific American* in October 1970 (Gardiner 1970). One of the first students in the Department was Julian Sullivan, originally an electrical engineer who later pursued undergraduate training in fine art at Middlesex Polytechnic (Hornsey School of Art) where he worked with the computer graphics pioneer John Vince. Sullivan was particularly interested in CAs and their potential. His work on boundary detection was adopted by the image processing researchers at UCL. Sullivan went on to join Briscoe on the staff of the computer studio where he worked until his death in 1982.

Amongst the students in the department who pursued an interest in the nascent field of A–Life was Steve Bell, who went on to complete a PhD with Edmonds at Loughborough. Bell produced several artworks using predator–prey A–Life models that were converted into graphics form using his 'Smallworld' software. Bell is now at the UK's National Centre for Computer Animation at Bournemouth.

I was a contemporary of Bell's at the Slade where I studied from 1977 to 1979. My primary interests were CAs and deterministic 'chaos'. Readers who wish to know more about my work can read about it in my essay in 'Stepping Stones in the Mist' (Brown 2002) or visit my website (Brown 2003).

The Australian art theorist Mitchell Whitelaw suggests that A–Life is a natural development of artistic practice throughout the 20th century (Whitelaw 2000). In particular he quotes the work of Paul Klee and Kasimir Malevich. Many artists have claimed that that their work has an independent life of its own and that the artwork 'tells' the artist when it's finished.

Those artists associated with the Slade's computer studio in the 1970s felt they were building upon the traditions of constructivism, systems art and conceptualism and that the computer was a 'natural' tool with which to continue this kind of work. Many of the more traditional artists associated with the studio and who did not use computers themselves agreed. We did not use the term 'Artificial Life' and would not especially have associated with the term as it was defined by Langton in the 1980s as a form of 'experimental biology' (Langton 1989). Our focus was more on procedure and process in their most general sense and, moreover, many artists actively resisted attempts to apply anthropomorphic interpretations to their productions. Nevertheless, references to life and physical and biological processes were often implicit in many of the works. Examples would include Conway's 'Game of Life' which had a major influence or my own time–based work 'Builder + Eater' (1977) where two concurrent processes dynamically competed for possession of a digital image.

Conclusion

The Experimental and Computing Department in the Postgraduate School of the Slade School of Art existed from 1972 to 1982. As I hope I have been able to demonstrate above, it was a dynamic focus for artists working with computers and especially for those concerned with computational and generative methodologies that, in the 1980s would become 'sanctified' by science with the classification 'Artificial Life'. My intention in writing this essay has been to put their endeavours on the record and ensure that this almost forgotten period of British art history is preserved.

Bibliography

Brown, P. (2002). 'Stepping Stones in the Mist', in Bentley, P.J. and Corne, D.W. (eds.). *Creative Evolutionary Systems*.: Morgan Kaufmann.

Brown, P. (2003). Personal website: www.paul-brown.com

Buchanan, B. G. and Shortliffe, E. H. (1984). *Rule–Based Expert Systems: The MYCIN Experiments of the Stanford Heuristic Programming Project*. Reading, MA: Addison–Wesley. See also: http://smi-web.stanford.edu/projects/history.html#MYCIN.

Cornock, S. and Edmonds, E.A. (1973). 'The Creative Process Where the Artist is Amplified or Superseded by the Computer'. *Leonardo* (6)1, pp. 11–15.

Edmond, E.A. (2003). 'Logics for Construction Generative Arts Systems'. *Digital Creativity*, (14)1, pp. 23–8.

Gardiner, M. (1970). 'Mathematical Recreations'. *Scientific American* 223, October, p. 120.

Kay, A. (1984). 'Computer Software'. *Scientific American Special Issue*, September, pp. 53–9.

Kurzweil, R. (2001). www.kurzweilcyberart.com

Langton, C.L. (1989). *Artificial Life*. Cambridge, MA: MIT Press.

LeWitt, Sol (1967). 'Paragraphs on Conceptual Art'. *Artforum* 5, Summer, p. 80.

MacGregor, B. (2002). 'Cybernetic Serendipity Revisited', in *Proceedings of Creativity and Cognition 2002*, ACM, pp. 11–13. Downloadable from: http://portal.acm.org/citation.cfm?doid=581710.581713

SAM video –http://members.lycos.co.uk/zivanovic/senster/videoclips/sam_long.mpg

Senstor video –http://members.lycos.co.uk/zivanovic/senster/videoclips/senster_long.mpg

Stravers, K. (2003). *The Evoluon*. http://home.iae.nl/users/pb0aia/evoluon/indexe.html .

Whitelaw M. (2001). 'The Abstract Organism: Towards a Prehistory for A–Life Art'. *Leonardo* (34)4, pp. 345–8.

Zivanovic, A. (2003). A website dedicated to the work of Edward Ihnatowicz: http://members.lycos.co.uk/zivanovic/senster/

5.2 The Interactivity of the Moving Document as the Diegetic Space of Consciousness

Clive Myer

> *The traditional academic position . . . maintains that the scientific validity of this Renaissance invention* (perspectiva artificialis) *is absolute and therefore timeless. Some of these theorists of perception fail to give due weight to the difference between our optic system and the way it is represented in images. They fail to notice that taking the eye as a model for vision is an ideology like any other, even though to us it seems more natural.*
>
> (Aumont 1997)

> 'document' –from the Latin documentum, 'lesson, proof'.
>
> (Concise Oxford Dictionary 1999)

The diegetic space of consciousness is the world we do not live in but still inhabit, the world we dream in while still awake, the street we walk down but do not see, the world we think we control but suspect we do not. This world that used to be gate–kept in days before the Quatrocentro by the key–masters of the monasteries and the goalkeepers of the world's churches doubling as guardians of the state, found its modern articulation in the post–human form of the visual document. This document, no longer devised through text nor understood through interpretation, no longer hand–mind produced as a sole object of the gaze (the painting) transmogrified both surprisingly and magically from the illustrative to the mimetic, from the pictorialised printed artifice to the ubiquitous photo–unimaginary document of the camera eye which narrated the world as assumed, needless of the orator as if to strip the fundamental principle of, say, the traditional Kabuki theatre of its on–stage narrator.

So what of the kinetic mutations of the last century?

For the purposes of this paper we will concentrate on the concept of documentary as the baton which has been passed on from still photography, which itself remains prevalent but is now a metadiscourse of the moving image, with an eye to the refracturing of the moving image in the digital post–cinematic culture. This triumvirate of concepts interactivity/moving document/diegetic consciousness alludes to a post–representational inquiry, in this instance, within a critical reappraisal of the notion of documentary film (now in crisis) with its grand allusion to and illusion of the real. It considers consciousness as never having escaped the primarily (but not only) prescribed world of the interior and exterior diegetic spaces of person–mind and film–mind.

The moving document (really any film whether documentary or fiction, silent or synchronised) demands recognition for its role in the misplacement of knowledge, if one is to engage with the process of linguistic/audio–visual consciousness in any epistemological sense. The old adage of the documentary as a window on the world has never really been resolved. The nature of the window itself has been addressed both ontologically, e.g. Vertov, 'I am a camera' and epistemologically, e.g. Comolli's reference to the similarity of the French word *objectif* meaning lens and objectivity, here in its particular company with documentary imagery and apparatus.

The nature of diegesis in this discourse is the one most important aspects of connectivity between the document, the decentred subject, interactivity and consciousness. The diegetic space of consciousness (space being an important concept, inasmuch as spatiality can be conceived as infinite and expansive as opposed to the finite expectancy of, in this instance, the documentary) differs from the notion of the unconscious or the subconscious in the sense that the signs and symbols of the psychoanalytic space operate as signals for another meaning. They stand in for some other 'truth'; they misguide the subject as an unwilling participant on an individual journey of subjectivity. Diegetic space, on the other hand, demands a collective space and acquiescent subjects. It operates seemingly visibly in the lived world and it incorporates all representation within its scope. It is the knowing and unknowing complexity of lived knowledge, of how we (mis)understand the world. Durkheim, the godfather of sociology and contemporary of Marx, may have been the first to recognise this in his writings on the 'conscience collectif' (Durkheim 1893[1]). Specifying religion as the major purveyor of state control, he pre–dated post–Marxist notions of ideological apparatuses by almost a century

The misused notion in film studies of diegesis as the on–screen or off–screen space of the narrative belies its complexity and its special place within the triangulated modernity of form, content and context. In its deeper sense, diegesis is the mental referent embedded in myth and locked between the subject in transition and the object of desire, the place occupied by the Other and inaccessible to the lived–in world. Its real home is in the represented world but its breathing apparatus exists in the lived–in world. We are the subjects of its gaze; we are its Other. We are the meaning produced by it and we go about our business warily.[2]

The coherency of any document relies on its underlying system, both physical and representational. Its historicity is transient and its value open to interpretation.

The apparatus of (pre–digital) still photography could be seen, in some ways, as equivalent to that of the earliest mechanisms of cinema: pre–camera movement and pre–editing. The mimetic space of the viewer lies between the still camera and the enlarger in a similar way that the physical place of the subject in early cinema lay within the apparent immobility of the cinematic camera/ projector (the same instrument used for taking the image and giving the image). In fact there is no stasis in cinema and little in the viewing of a photograph. Both are moving documents, travelling images. In gallery photographic exhibitions the spectator travels past the image and in home/personal photography the document is passed around, whereas in film the image travels past the spectator, both overhead and vertically on screen. Ironically, the moving document could also be interpreted as emotionally affective pathos where the document entraps the spectator before devouring its prey, its ability to collectively consume its audience with predetermined emotional switches. We exchange personal freedom for emotional intensity. Included in the package, as an unregistered bonus, is the ideological and political intensity of the coming–into–being or, in Deleuzian terms, the act of becoming, through the lack of familiarity with our own Other in confrontation with the object Other.[3]

The viewer is an active participant in these processes or, more accurately, an *interactive* participator. There is a 'lock–on' (sometimes as precise as military 'smart' machinery) and a 'lock–in' (as sociable as in a pub). The feel–good factor is payback for the interactivity between the exchange values of the economic system. In the case of cinema, the equivalent is the active engagement between the gaze of the cinema screen and the gaze of the viewer. The viewer, caught between the gaze of the projector and the gaze of the screen, is known from behind and from the front almost simultaneously. There also exists an interactive virtual link through the Deleuzian sense of time –interpreted by Jaques Aumont as one that 'almost allows us to *perceive* time' when we move through different levels intertextually –past, present and future narrative time which exists in an extra–real sphere that resides outside of experienced functional time (Aumont 1997 'The Role of the Apparatus' p. 130).

Through the parameters of the discourse set up in this paper, one is bound to return to the sociological implications of western society's cinematic consciousness, from its contemporary society of 1895 to the ponderous and seemingly ahistorical context of today's post–grand narrative. Ella Shohat and Robert Stam describe the relationship between national self–consciousness and the context of the invention of cinema:

> *The nation of course is not a desiring person but a fictive unity imposed on an aggregate of individuals, yet national histories are presented as if they displayed the continuity of the subject writ–large.*

(Shohat and Stam 2002 p. 118[4])

They describe the role cinema took at its genesis of the 'being–there' of the camera to collect and disseminate ideas of empire, to develop and sophisticate the diegetic space of consciousness upon which, they argue, were built the great cinematic genres of our time.

Today, the smart bomb may be conceived as inheritor of the moving document's role of imperial consciousness. Associated with our opening quotation from Jaques Aumont, perspective's recentring takes into account the view or gaze of the bomb, which becomes vertical perspective. It takes us in full circle from the grand philosophical project of the proof of God and Durkheim's attempt to understand the role of religion in society to the socialised diegetic post–representational project of the interiorised role of modern physical and ideological 'techno–God' consciousness. In other words, we have shifted from the initial all–powerful vertical view of God–on–high creator/destroyer to the enlightened all–powerful Renaissance Man's horizontal view of the world to the all–powerful vertical view of the world, displacing the omnipotence of the creator (the feared 'Other') with the omnipotence of man and machine.

They conclude by stating:

> *If postmodernisim has spread the telematic feel of First World media around the world, in sum, it has hardly deconstructed the relations of power that marginalize, devalue, and time and time again massacre otherised peoples and cultures . . .Within the Gulf War, the fact of mass death itself, the radical discontinuity between the living and the dead, reveals the limitations of a world seen only through the prism of the simulacrum,*

(ibid p. 142)

The symbolic place of the viewer–in–the–world gives birth to the symbolic place of the viewer–in–the–cinema, in turn giving birth again to the symbolic place of the viewer–in–the–world and so on. This develops in complexity with the loss of the grand narrative of modernism. There is, theoretically, no longer one world in danger of splitting in two but a series of worlds with parallel dimensions of cinematic and live–in places that become science/fiction worlds (not science fiction but its Other), inhabited by both mind and body. This appearance of pluralism produces a diegetic fragmentation through cinetelematic apparatuses not that far from fragmentation used by pornography. The resultant effect may well become not one of false consciousness but fractured consciousness, pacification through tokenism, deep social frustration. Waiting in the wings of this liquid, mercury–like, theatrical atomisation is post–globalisation, eager to rehomogenise difference but unsure of its omnipotent responsibility.

Where now is the voice, the narrator in documentary? Will the return of the Kabukiesque –not as off–screen voice of God, as film studies may have analogised, but as consciousness itself materialised, as interiorised storyteller compromised impossibly without fiction, born of contradiction between truth and depiction between the Real (Other) and the real (other) and the reel (object–cause). Filmed material becomes the object of desire; the unachievable that sends the viewer's drives (ego, sexual, life, death) in endless circles.

Bibliography

Aumont, Jacques (1997). *The Image*. London: British Film Institute.

Balibar, E. and Wallerstein, I. (1991). *Race, Nation, Class: Ambiguous Identities*. London: Verso.

Concise Oxford Dictionary (1999).

Deleuze, G. (2001) *Cinema 1 –The Movement Image*. London: Athlone Press.

Shohat, E. and Stam, R. (2002). 'The Imperial Imaginary', in Askew, K. and Wilk, R. *The Anthropology of Media*. Oxford: Blackwell.

Notes

1. Durkheim articulates social beliefs and values in terms of volume, intensity, rigidity and content.
2. Deleuze refers to Eisenstein's concepts of montage as having created a 'transforming form': 'Eisenstein's transforming form often demands a more complex circuit: the transformation is indirect and thus all the more effective' (Deleuze 2001 p. 181).
3. For example, in both the Bruce Willis 'movie' (sic) *Armageddon* 1998 and a TV episode of 24 screened in the UK in June 2003, identical scenes between a daughter seeking her father and his imminent death share the same structured and coded use of predictability of audience tears.
4. This quotation itself refers to a more concise passage from Etienne Balibar: 'The histories of nations are presented to us in the form of a narrative which attributes to these entities the continuity of a subject' (Balibar and Wallerstein 1991 p. 86).

5.3 Assimilating Consciousness: Strategies in Photographic Practice

Jane Tormey

A model of reframed consciousness can be explored within the context of contemporary art photography. If philosophical/theoretical discourse concerning consciousness affects attitudes to 'reality', then photographic practices visibly reflect changes in those attitudes and present them in current artistic preoccupation. The interrelationship between photographic strategies and consciousness might be described as an aspect of 'post–modernism', which Jean–Francois Lyotard (1984) defines as a radically altered understanding of reality. In the context of art, he explains that successive reactions to knowledge test the rules of existing aesthetic practice, provoking successive innovations –resulting in a series of anti–aesthetics. Post–modernism is a recurring state, where 'beliefs are shattered', where other realities are invented, a period of attitudinal adjustment, coming before (ante) assimilation. Typical reaction and assimilation now centres around three premises brought to the fore in current thinking –an understanding of the image as not real (Baudrillard 1997b), uncertainty provoked by Jacques Derrida's (1973) repositioning and an instability arising from loss of the author (Barthes 1977). The profound influence of these ideas on information, on knowledge, on our understanding and ultimately our aesthetic assumptions (consciousness) can be seem in examples of photography –in strategies that dictate the images. Methods of assimilation echo qualities, such as non–determinacy, which circumvent the impossible task of making definitive photographic statements –assuming methods of avoidance/obliqueness/b landness/ordinariness/ar tificiality/atte nuation/contradiction.

Vilem Flusser describes our relationship to the world as dislocated owing to the impact of the image. Our understanding, mediated by photographs, reflects a 'second–order magical consciousness' (Flusser 2000 p. 17) that has assimilated the inauthenticity of 'reality'. In his discussion of simulacra, Baudrillard identifies the problem for representation that results from a reality constructed by the image and where we can no longer distinguish between real and imaginary, original and copy, surface and depth. He assumes that the real is lost and advocates that we 'give up representation'. 'Hyperreality' and 'aesthetic illusion' have repositioned photographs as equivocal documents rather than a reflection of reality. Our grasp of reality is thus key in the evolution of photographic strategy. Because the indexical quality of photographs has led us to believe that what is presented is real, so photographs easily disguise or contradict, bear no relation to reality at all and can be entirely fabricated. In his writing on photography specifically, Baudrillard proposes an un–definition of what is real (Baudrillard 1997a). If the central tenet that there is an authentic truth to be found through endeavour is no longer viable, then it becomes meaningless to pursue or represent reality and logically one should actively abandon the attempt. He challenges the photographer to 'disappear', to relinquish interpretation, so that the depicted subject can speak. He proposes that the author no longer searches for 'essence' or universal quality and accepts rather an authenticity mediated by the reader –a disappearing of the photographer as subject. He empowers the reader and the subject depicted but not the author's constructed idea of the subject. What Baudrillard proposes and what much contemporary work adopts is a lack of control or intentional meaning, allowing meaning to assert itself by way of the insignificant and ordinary in the image. He turns the role of the photographer inside out, uprooting the central role of the author as expressing herself. In the face of simulacral confusion, he promotes an oblique re–emergence of the real. Decentred, less complete, less focused.

The premise of uncertainty and contradiction, allowed by Derrida's deconstruction and critique of logocentricism, is central to a new kind of 'disinterestedness' apparent in practice. Artists are seen to be denying received knowledge/aesthetic convention in a way that distances them from the 'logocentric' and are adopting instead more circumspect and subtle strategies of non–linear narrative, positional uncertainty, incidental view and a reliance on instinct and happenstance. In *Droit de Regards* (Derrida 1989) Derrida examines a series of photographs and explores 'interminable' narrative, contradicting and challenging our 'desire for stories' and resolution. His analysis allows every detail to have significance, steering us away from a definitive account, denying us the certainty of closure, demonstrating methods of looking and understanding through his questioning of implicit interpretation. He celebrates the lack of any one underlying meaning, the uncertainty, the multiplicity and the possibility of non–oppositional contradiction. Such lessons in the reading of images as non–definitive and contradictory are paralleled by photographers who avoid the obviously 'meaningful' and who approach the image obliquely, presenting imploded depictions of consciousness, of indiscriminate ordinariness in their subversion of traditional modes such as portrait and landscape. Issues of authorship and authenticity effected in photographic work have a more fundamental influence than the oft–cited examples of post–modern photography such as the literal artificiality of depicting false identities (Cindy Sherman), false origin (Sherrie Levine), false reality (Andreas Gursky). More profound subversions of a traditional photographic perspective extend logically by fabricating the appearance of reality as the ordinary, actively seeking a denial of the author as subject and using artificial strategies to effect a disruption of roles or to present a centre–less view and irrelevant content. Artists deliberately adopt strategies to do this, by eliminating themselves and by avoiding the expressive or by being as inauthentic as possible. The attenuated work, for example, of Beat Streuli and Thomas Ruff abandons expressive photographic intervention or the search for expression in the depicted subject and reflects aspects of Baudrillard's 'disappearing subject'. They are both non–expressive and expressionless, having effected a subversion of direct expression. The extreme deliberateness of Bettina von Zwehl and Ulf Lundin contrive its total rejection. Ruff verifies Baudrillard's impossible realm of reality and provides an example of non–representation/non–meaning. His authenticity lies with the primacy of the image over the photographic event and his determination not to succumb to the illusion of being able to represent. He presents a kind of 'second order' reality, the 'thingness' of the photograph rather than a person –the person as an abstraction.

A shift in attitude as a consequence of the 'death of the author' and the understanding that 'objectivity' is an illusion (Flusser 2000 p. 15) moves towards either an interdependence of subjects (photographer and photographed), a confused intimacy or an avoidance of involvement altogether. Not only has Barthes diminished the role of the author but also in his subjective exploration of the photograph that confuses theory with emotion (*Camera Lucida*), he has encouraged a subjective exchange that confounds objectivity. Photographers no longer have the certainty of their own authority or that of objective vision. Self–consciously aware of this disrupted authorship, they have adopted overt methods that divert any accusation of authorial inspiration. In Thomas Struth's case, he purposefully refuses to allow his own subjectivity, his 'idea', to dominate, giving no direction beyond determining the extent of the frame. Beat Streuli and Philip Lorca di Corcia use photographic devices, which remove them entirely from their subjects and set up a theatre of appearance. They disallow any attempt to reveal the 'real' nature behind the mask, presenting the appearance over anything else, without intentional projection, structure or meaning, as far as this is possible. They present in their ordinariness and non–event an ontology of boredom, an indifference of seeing, and relinquish any inspirational or directorial determination of the image. These images present the contradiction of an apparent translucence of appearance and an opaqueness of subject matter and consequent emptiness. They give us appearances that ultimately obscure.

Both Lundin's *Pictures of a Family* and Annelies Strba's *Shades of Time* give us visions of family life in parenthesis, Lundin's via avoidance and Strba's via intimacy. We associate photographs of family with event and interaction. Lundin's images document non–event, highlighting relationships by showing us the lack of interaction. They focus on remnants: what is normally left over and discarded. They look sideways at a life, indicating the incidental and the ordinary, what is not said. They present us with a fragmentary, slanted view of what we are looking at, an emergent awareness of individuals. Strba closes her eyes when pressing the shutter, not seeing, 'disappearing', denying the intention of the 'photographic eye'. It is a method that relinquishes power and a substantial part of the traditional position as photographer, by not preparing images for the viewer. Both assume the

Figure 1: Ulf Lundin, from the series Pictures of a Family, 1996

validity of photographic series, replacing the dualism of essential being and appearance, and emphasise a process where there is no ultimate end; where all one can find is a series of manifestations; where beings change and will present themselves differently at different times. Dialogic imagery is dependent on singular individuality and the specific detail of context, approaching a sort of meaninglessness. By giving us the ordinary rather than the extraordinary (literally outside the norm) bereft of a directed expression, dilemma or passion, their work begins to undermine the presumption that the photographer has something to say or find; it undermines the search for resolution and significance through metaphoric reference, a shared, greater meaning. These are more discursive, are more overtly inter–subjective, unrehearsed, uncontrolled methods. They lead us away from the presumption of 'presence' to a more open field.

These images present little allusion to other than what is there, little scope to render the subjects as anything beyond themselves. But despite the eschewal of the 'captured' moment, despite a kind of metaphoric minimalism, as Derrida demonstrates, the 'metaphotographic event' is impossible to avoid: what went before, what comes after, what is imagined, metaphor, metonym is held in each of these ordinary eventless moments. In reading these images, even the most simple statement, such as Sonja's right hand hovering over the glass, leads us elsewhere, to our imagination, penetrating 'the abyss of these metonymies'. The viewer is thus assigned a speaking role that can speculate and position, where 'there is reversibility, irreversibility, diachrony and simultaneity'. 'The fragment should remain 'discreet' (Derrida 1989) if it is to retain any potency. It is neither central to the image, nor significant in itself, even irrelevant, but without the glass of water the image would be either meaningless or more meaningful. Baudrillard has suggested that 'poetic order requires that the event should not exactly take place' (Baudrillard 1999) and advocates that the activity of taking the photograph itself be pivotal, be kept crude and uncontrolled rather than the prospect of the resulting image being in the forefront, thus avoiding 'photography that is aestheticised, calculated and composed' (Baudrillard 1997). Recognising that an accidental or unassuming image can equally be eloquent leaves the

Figure 2: Annelies Strba, Sonja with a Glass, 1991

photographer with the ironic possibility that artifice or lack of artifice may be equally deceptive or meaningful. What is 'real' need no longer be elevated or even made beautiful. Much contemporary work embraces this contradiction by deliberately looking crude or abrupt (Boris Mikhailov) or deliberately incidental (Ulf Lundin) or deliberately chaotic (Nick Waplington).

A 'traditional' or modernist photographic aesthetic has been amplified by the notion of the 'photographer's eye', demanding the specialness of the author, control and vision in a representation of the event and moment –a revisioning of the everyday. Not only has photography now been absorbed as a legitimate mode of artistic production but the assimilation of the snapshot as a genre disrupts the seriousness of photography as it does not conform to the ethic of artistic distance, a prerequisite for 'objective vision'. Liberated by the snapshot, a democratic seeing has been assimilated into our aesthetic and directorial subjective expression has been replaced with a subjectivity that can confuse intimate and professional roles, can be careless and ugly and approach bad taste.

Lyotard states that the project of modernity as 'the realisation of universality' has been relinquished. One can see this destruction in process, in the avoidance of universal appeal and a preference for particularity and insignificance. Western philosophy is habitually driven by the goal of finding or achieving a unifying principle and it follows that what has been seen as 'good' art pulls things together in some sort of synthesis. Western aesthetic has assumed an 'objective vision'. If aesthetic trend is reactive, then we are reacting now with the adoption of the unremarkable and the awful as 'good', relinquishing formalism and the supremacy of the intentional 'photographer's eye'. A deliberately crude realism is one logical step to dismantle this hierarchy. 'Post–modern photography' is typically and 'essentially' goal–less, disregarding the long held assumption to unite and to complete, together with the modernist assumption of photography that must define or mythologize. The construction of a changing (anti) aesthetic is not reliant on universal certainties or knowledge, is less certain –is malleable rather than 'fixed or permanent' (Moxey 1999), moving towards forms of local specific knowledge that can be seen in choices of subject matter.

An aesthetic has arisen, which assumes methods accordingly, avoiding interaction and expression –hiding, closing eyes. There is now a distrust of the author and the image as representing any sort of reality and an insistence instead that the image is constructed by the reader, culture and history. There is, in effect, almost an abdication of authorial responsibility, an undercurrent of denial –an aesthetic of 'without'. Denying an underlying truth or essence or anything to be revealed or told, there is no ultimate description, no definitive image, no 'moment'. Contemporary photography has abandoned reality and adopted an 'idea of reality' and is abstract in the sense that it focuses on the idea rather than the form and the substance –the frame and the composition –photographic 'aesthetic illusion'. It displays an imperative to avoid direction and to avoid definition –typical characteristics of post–structural/post–modern texts, of interrupting and subverting traditional forms. The photographer knows that photographs are not real (like Baudrillard), 'mixes' with the subject (like Barthes), interferes and obscures (like Derrida).

Bibliography

Barthes, R. (1977/1968). 'The Death of the Author', in Heath, S. (ed.). *Image, Music, Text*. London: Fontana Press, pp. 142–8.

Baudrillard, J. (1997a). 'The Art of Disappearance', in Zurbrugg, N. (ed.), *Art & Artefact*. London: Sage, pp. 28–31.

Baudrillard, J. (1997b). 'Objects, Images and the Possibilities of Aesthetic Illusion', in Zurbrugg, N. (ed.), *Art & Artefact*. London: Sage, pp. 7–18.

Baudrillard, J. (1997c). 'The Ecstasy of Photography', in Zurbrugg, N. (ed.), *Art & Artefact*. London: Sage, pp. 32–42.

Baudrillard, J. (1999). 'It is the Object that Thinks Us', in Wiebel, P. (ed.). *Photographies 1985–1998*. Hatje Cantz.

Derrida, J. (1973). 'Difference', in *Speech and Phenomenon*. Allison, D.B. (trans.) Evanston, IL: Northwestern University Press. Originally published in the *Bulletin de la Societe francais de philosophie*, LXII, no. 3 (July–September1968), pp. 73–101.

Derrida, J. (1989). 'Right of Inspection'. Wills, D. (trans.) 'Droits de Regards', in *Art & Text*, 32, Autumn 1989, pp. 10–95.

Flusser, V. (2000/1983). *Towards a Philosophy of Photography*. London: Reaktion Books.

Lyotard, J.F. (1984). *The Post–Modern Condition: A Report on Knowledge*. Bennington, G. and Massumi, B. (trans.). Minneapolis: University of Minnesota.

Moxey, K. (1999). 'The History of Art after the Death of the "Death of the Subject"', in *[In [ ] Visible Culture*, www.rochester.edu.in_visible_culture/issue1/moxey/moxey/html. Accessed 1 February 2003.

5.4 Simultaneity, Theatre and Consciousness

Daniel Meyer-Dinkgräfe

All theatre has to take *time* into consideration. On a practical level, the question arises as to how long a performance should last. How many hours can a given audience be expected to pay attention to the theatre event? Originally, *Noh* performances in Japan consisted of several plays in a row, with the comedy form of *kyogen* in between. In India, performances lasting several days were known. In the West today, attention spans are bemoaned to become shorter and shorter, owing to television soap operas and thus plays wishing to be commercially viable have to be adapted and be no more than two to two-and-a-half hours including interval(s). Subsidised companies, such as the Royal Shakespeare Company or the Royal National Theatre, London, may exceed this limit, as may occasional experimental productions by acknowledged stars of the theatre, such as Peter Brook's nine-hour *Mahabharata* or Peter Stein's 21-hour production of Goethe's *Faust*.

Time is, of course, not limited to the duration of a production and the commercial, socio-cultural and psychological issues associated with it. Time is an important feature of both drama (the literary text) and the performance in the theatre. Plots may progress in a linear fashion, starting at point in time (A) and moving on steadily via points (B) and (C) to the end of play at point in time (D). The time it takes spectators to watch the play is the time that is suggested to have passed in the fictional reality of the play. Other plays would still fit the category of linear but some stages in the development are not shown on stage, rather occurring off-stage, and may be reported as past events on stage. Plays may reverse the sequence of events over time, beginning at a chosen point and moving back from it (as in Pinter's *Betrayal*). Some contemporary dramatists have explicitly experimented with time: in *Noises Off*, Michael Frayn juxtaposes events on stage with simultaneous events backstage, presenting us first with the on-stage scene (the play within the play), then the backstage events while the scene we had seen before on-stage takes place off-stage. The same span of time (the presentation of a scene from the play within the play on the fictional stage on the real stage) is shown twice, from different perspectives. In one of Alan Ayckbourn's more recent plays at the Royal National Theatre, London, *House and Garden* (2000), the same fictional time span is presented simultaneously in two of the three theatre spaces in the RNT (Olivier and Lyttleton) by the same cast, presenting indoor and outdoor perspectives.

In all those cases attempts are made to convey the intricacies of time. However, hard as they try, Frayn and Ayckbourn have not managed to achieve the impression of complete simultaneity. Spectators intellectually *know* that this is a clever device. When they are watching the first part of Ayckbourn's *House and Garden* in the

Olivier, they *know* that the situation they have seen is continued next door in the Lyttleton. Once they have seen the first part and now proceed to the Lyttleton, they can match what they see now with their memories of the performance in the Olivier. However, this matching activity is also intellectually mediated, not immediate. In Frayn's *Noises Off*, matching the events of the scene from the play within the play on stage with the events backstage is made easier by hearing at least some of the on–stage text while the same scene is repeated from the backstage perspective. However, true simultaneity is not achieved.

A look at Vedanta philosophy can elucidate the specific effect of simultaneity on the stage and show its psychological significance for the actors and the audience alike. Vedanta philosophy is concerned predominantly with consciousness, which, as subjective monism, is located at the base of all unmanifest and manifest creation. Time fits in with this approach. The western mindset, in its aim for scientific objectivity, associates time with a sequential ordering of events, studied in the discipline of history. According to Vedanta, the emphasis of history is on the importance of events, not on chronology, because of the conceptualisation of time as eternal. The following passage provides a rather mind–boggling account of how time is conceptualised in Indian philosophy.

> *The eternity of the eternal life of absolute Being is conceived in terms of innumerable lives of the Divine Mother, a single one of whose lives encompasses a thousand life spans of Lord Shiva. One life of Lord Shiva covers the time of a thousand life spans of Lord Vishnu. One life of Lord Vishnu equals the duration of a thousand life spans of Brahma, the Creator. A single life span of Brahma is conceived in terms of one hundred years of Brahma; each year of Brahma comprises 12 months of Brahma, and each month comprises thirty days of Brahma. One day of Brahma is called a Kalpa. One Kalpa is equal to the time of fourteen Manus. The time of one Manu is called a Manvantara. One Manvantara equals seventy–one Chaturyugis. One Chaturyugi comprises the total span of four Yugas, i.e. Sat–yuga, Treta–yuga, Dvapara–yuga, and Kali–yuga. The span of the Yugas is conceived in terms of the duration of Sat–yuga. Thus the span of Treta–yuga is equal to three quarters of that of Sat–yuga; the span of Dvapara–yuga is half of that of Sat–yuga, and the span of Kali–yuga a quarter that of Sat–yuga. The span of Kali–yuga equals 432,000 years of man's life*

> (Maharishi Mahesh Yogi 1969).

Clearly, any attempt at chronology, given this conceptualisation of time, would be counter–productive, as would any attempt to grasp this concept of time intellectually. Vedanta does not expect us to do this. Instead, it argues that human beings may experience the infinity of time (and space) in their own consciousness, in a specific state termed pure consciousness. Vedanta philosophy not only provides the conceptual framework but offers, in addition, practical techniques said to enable such experiences.

Current cognitive psychology is aware of the need to develop a 'taxonomy or architecture of mind that accounts for the structure and dynamics of a wide range of mental processes engaged in by an adult knower' (Alexander 1990). In a publication on higher stages of human development, the authors introduce the most ancient and extensive description of 'developmental transformations in adulthood' (ibid) as found in the Vedic tradition of India. This has recently been formulated by Maharishi Mahesh Yogi as Vedic Psychology. Vedic Psychology 'proposes . . . an architecture of increasingly abstract, functionally integrated faculties or 'levels of mind'' (ibid). This hierarchy is one that ranges from gross to subtle, from highly active to settled, from concrete to abstract and from diversified to unified (ibid). Vedic Psychology specifies the following levels of mind:

> *The faculty of action and the senses, desire, the thinking mind, the discriminating intellect, feeling and intuition, and the individual ego . . . (Note that mind is used in two ways in Vedic psychology. It refers to the overall multilevel functioning of consciousness as well as to the specific level of thinking [apprehending and comparing] within that overall structure.) According to this theory, underlying the subtlest level of the individual knower and transcendental to it is the Self, an abstract, silent, completely unified field of consciousness, identified as the self-sufficient source of all mental processes.*

(ibid)

Repeated exposure to pure consciousness, achievable, for example, through meditation, in alternation with activity of any kind will lead, over time, to the development of even further stages of human development. Based on the proposition to regard pure consciousness as a fourth state of consciousness next to the commonly known and experienced ones (waking, dreaming and sleeping), Vedic Psychology describes three further stages of consciousness development. Termed 'cosmic consciousness', a fifth state of consciousness is characterised by the coexistence of waking or dreaming or sleeping and pure consciousness. In cosmic consciousness, the level of pure consciousness, which is never overshadowed in daily life by the activities and experiences of the expressed mind, becomes a 'stable internal frame of reference from which changing phases of sleep, dreaming, and waking life are silently *witnessed* or observed' (Alexander and Boyer 1989). The next stage of development, according to Vedic Psychology, is called 'refined cosmic consciousness'. In cosmic consciousness the field of pure consciousness is permanently experienced together with waking or dreaming or sleeping. This level of functioning is maintained in refined cosmic consciousness and 'combined with the maximum value of perception of the environment.

Perception and feeling reach their most sublime level' (ibid). The final level of human development according to Vedic Psychology is called 'unity consciousness'. In this state of consciousness 'the highest value of self–referral is experienced'(ibid). The field of pure consciousness is directly perceived as located at every point in creation, and thus 'every point in creation is raised to the . . . status' of pure consciousness (ibid). 'The gap between the relative and absolute aspects of life . . . is fully eliminated' (ibid). The experiencer experiences himself and his entire environment in terms of his own nature, which he experiences to be pure consciousness. In all three higher states of consciousness as conceptualised in Vedic Psychology, pure consciousness is a permanent aspect of daily experience and with it is infinity of time and space as the basis of the obvious limitations of those dimensions in manifest creation.

Now that I have established some insights of Indian philosophy in relation to time and consciousness, I can return to a reassessment of that specific function of time in the theatre, simultaneity. Usually plays are directed in such a way that there is always only one scene on the stage. Even Frayn's and Ayckbourn's attempts at simultaneity do not break this rule. The spectator's attention is allowed to focus fully on that scene. In that scene major characters will carry the scene, while other performers better be in the background physically and emotionally so as not to upstage those at the scene's centre. In contrast, for example, take the production of Mozart's opera *Marriage of Figaro* by David Freeman and the Opera Factory Zurich. Life of the house of Count Almaviva is shown throughout, breaking the boundaries of ordinary opera direction. The traditional rule is to have only the characters on the stage who have to sing something at that time. Not so in Freeman's *Figaro*. Life in the house goes on. The threads of the story come through the house. The focus of the scene is on the singers but other characters go about their respective business at the same time. A few examples should illustrate this: during the overture, Don Curzio enters, sits down at a table upstage right and starts writing. Soon he is joined by Basilio. At the same time, Antonio brings parts of a wooden bed into the small area designated as Figaro's and Susanna's chamber downstage left. After that, Antonio moves to an area centre stage right that represents the garden, indicated by flowers, and starts preparing a beautiful flowerbed. Barbarina, Susanna and two maids are busy in Figaro's and Susanna's chamber, cleaning the floor. Cherubino joins them: he enjoys female company. Before the end of the overture, Figaro arrives, makes all others except Susanna leave, so that they are now ready for their opening scene. Meanwhile, Bartolo and Marcelline have also appeared, downstage right, and during Figaro's and Susanna's scene one of the maids brings hot water for a footbath and a camomile–steambath against Marcellina's cold. Basilio moves to 'his area', indicated by a music stand, and starts composing. The maids start washing and wringing linen in the background, Basilio eavesdrops on Susanna's and Marcellina's quarrel.

Such a liveliness of parallel action is kept up throughout the production. At the same time, perception habits of the audience are provoked. Whereas the ordinary theatre experience means focusing on one central element on the stage, in Freeman's production of *Figaro* a flood of visual input reaches the spectator. All elements of input are interesting and make much sense because they are logical elements of the interpretation that Freeman provides. The audience has to learn to focus on the main element, which is provided by the music: they have to focus on the singers while at the same time allowing the other input not to distract but to enrich the insights gained from focusing.

Freeman carried his use of simultaneity even further in his adaptation of Mallory's *Morte d'Arthur*. The production was presented in two parts. The first half of Part I and the second half of Part II were presented in the traditional space of the Lyric Hammersmith, London. For the second half of Part I and the first half of Part II, the audience assembled in nearby Hammersmith Church. The space was empty (the pews had been removed) and the action took place simultaneously around five mobile pageants. Spectators had the choice to follow one storyline or to shift between the pageants or to stay somewhere in the space and just take in what happened to make its way towards their perception.

Simultaneity of space and time is a characteristic of pure consciousness. on this level of creation, past, present and future coexist. If a form of theatre forces the human mind to engage in the experience of simultaneity, it trains it in functioning from that deep level. Repeated exposure to such theatre stimuli may serve in parallel to repeated exposure to pure consciousness in meditative techniques. Theatre, understood and practised in this way, may thus well serve as a means of developing higher states of consciousness.

Bibliography
Alexander, C. N. et al. (1990). 'Growth of Higher Stages of Consciousness: Maharishi's Vedic Psychology of Human Development', in Alexander, C.N. and Langer, E.J. (eds.). *Higher Stages of Human Development: Perspectives on Human Growth*. New York, Oxford: Oxford University Press, pp. 286–341.

Alexander, C. N. and Boyer, R. W. (1989). 'Seven States of Consciousness: Unfolding the Full Potential of the Cosmic Psyche in Individual Life through Maharishi's Vedic Psychology'. *Modern Science and Vedic Science*, vol. 2, no. 4, pp. 324–71.

Maharishi Mahesh Yogi (1969). *On the Bhagavad–Gita. A New Translation and Commentary*. Harmondsworth: Penguin, ch. 1–6.

5.5 Cinematic Soteriology: Darshanic Effects in the Tamil Bakthi Films

Niranjan Rajah

To our Victorious Muruha Perumal[1] of Vayaloor,

All Praise!

Just as I am presently recounting and revealing to you

Six miraculous earthly manifestations of our great Arumuga Perumal;

Poets have sung the praises of Skanda Perumal,

Thereby enjoining in his Bliss;

Sculptors have refined the image of the Cenpaha Perumal,

Therein articulating his adoration.

In the same way this worthy group is assembled here,

To capture on film and to propagate the Lord's bakthi by cinematic means.

May this good work endure,

May it increase!

Tirumuruga Kirupananda Variyar[2]

The pneumatology and psychology of mythic and religious symbols

According to the traditionalist school of 20th century religious and philosophical studies, symbolisations of the divine or the 'world of gods' are metaphysical statements of an essential or transcendent reality –that of the *Philosophia Perennis Universalis*. In this context, Ananda Coomaraswamy (1977a) has explained that the myth is always true as it is 'a symbol, a representation . . . of the reality that underlies all fact but never itself becomes a fact'. For Coomaraswamy, symbolism is a hieratic and metaphysical language. A religious symbol states a metaphysical truth. It is a given –a revelation handed down from time immemorial. In another approach that draws from both Mircea Eliade and Carl Jung, Joseph Campbell suggests that mythology is 'a production of the human imagination, which is moved by the energies of the organs of the body . . . These are the same in human beings all over the world and this is the basis for the archetypology of myth' (Coomaraswamy 1977a). These views, the one linguistic/metaphysical /pneumatological[3] and the other biological/humanistic/psychological, while being in some ways at odds with one another, complement each other in framing the notion of soteriology.[4]

The soteriological instrumentality of sacred art

In pneumatological terms, the soteriology imperative is the attainment of 'gnosis' and in psychological terms the attainment of a shift from ordinary consciousness to what Campbell has called 'Transcendent Energy Consciousness'. This gnosis or shift in consciousness can also be understood as a transcendency —a transfer of attention from the indeterminate perceptual data of the surface appearances of the world, to its essential truth or underlying reality (Guenther 1975). Titus Burckhardt (1976) has defined this sacred art as the formal manifestation of the spiritual vision of a given religious system. It is by engagement with their respective 'formal manifestations' that the soteriological 'visions' of the various religions can most readily be attained. Indeed, it is the form and temporality of sacred art that give access to the formless and timeless realm of the divine.

Darshan and the emotionalist soteriology of *bakthi*

Darshan is literally 'seeing' and refers to gaining knowledge or consciousness of the divine via active visual perception (Eck 1998). This involves the worshipper seeing or beholding a sacred image with his or her eyes and actualising this image within the heart, therein being seen by the divinity in return (Coomaraswamy 1935). In the *bakthi* or devotional path of Hinduism, this actualisation is not approached in the quietude of contemplation. Instead, in *bakthi* woship the inner image of *darshan* is the catalyst for an effusion of love by which the desired transcendency is achieved. Central in this approach is the conversion of the aesthetic emotions awakened by the specific image of the *Devata* into the unbounded love that is, ultimately, the vehicle of *bakthi* soteriology. The three main systems of contemporary Hinduism all have their *bakthi* streams. The patrons of *bakthi* Hinduism are the poet saints and holy devotees —the *Saiva Nayanars, Vishnava Alvars, Shakta Baktars*. The most important of these are the poets who in medieval times composed and sang devotional hymns in praise of the specific divinities enshrined in the temples of south India. It is the life stories of these medieval saints that, along with mythological and miraculous tales, form the subject for the melodramatic musical productions of Tamil devotional cinema.

Folk art and modernity

In Coomaraswamy's understanding, myth, folklore, symbolism, the social order and the whole realm of traditional arts and crafts are 'all of one piece'

(Coomaraswamy 1977a). In defence of the non–canonical traditions he has said that, 'The material of folklore should not be distinguished from myth' but 'What we owe to the people . . . is not their lore, but its faithful transmission and preservation (Coomaraswamy 1977b). What is required of today's popular arts, then, is the faithful transmission of the received mythic symbols. However, even when 20th century art forms have met this criterion they have been rejected by the traditionalists because of their unconventional naturalism and sentimentality. In this regard, this paper proposes that, with the hindsight of the 21st century, the impact of modernity on the folk traditions must be accounted for in a more inclusive manner. If the hallmark of modernism in the West is the adoption of abstraction, modernity in the East has been marked by the encroachment of realism. This is particularly so in the popular traditions. Many transformations were unwitting assimilations made in the course of adjustments to the cultural and technological developments of modernisation.

Naturalism and visual piety

The paintings of Raja Ravi Varma are a case in point. His work bridges the traditional Indian painting of the Thanjavoor school and western academic realism. More significantly, as Ravi Varma influenced the pioneering filmmaker Dadasaheb Phalke,[5] it is an important point of departure for the Indian cinema (Kapur 2000a). In 1891 Ravi Varma founded the Ravi–Vijaya Press which was one of the earliest to produce colour lithographs in India. Varma's affordable oleographs of naturalistic, even sentimental, renditions of Hindu deities[6] quickly adorned the altars of many Indian homes. These images are, to this day, used extensively in domestic worship and have even found their place within temple ritual. In this light, it seems appropriate to temper the formal rigour of 'sacred art' with a term from post–modern discourse on popular religion –'visual piety'. This term refers to a broader notion of the visual formation and practice of religious belief. It is open enough to include canonical sacred art as well as all forms of popular religious art (Morgan 1998). It is proposed here that the melodramatic productions of *bakthi* cinema, like the sentimental Ravi Varma lithographs, are objects of visual piety and, in this regard, present darshanic effects and soteriological potential.

Cinema as technogically mediated ritual

It has been argued that the structure of viewing in popular Indian cinema is, as a whole, darshanic. As opposed to the isolated, individualised voyeurism of the unseen viewer in popular western cinema, in Indian cinema the object engagingly gives itself to be seen by the viewers collectively (Prasad 1998). It has also been argued that religious *darshan* and filmic diegesis are analogous modes of active or interactive seeing (Lovejoy and Jacob 1999). This paper proposes that, far more than being a useful metaphor, analogue, analytical devise or structural model in a secular reading, *darshan* coalesces with filmic diegesis to produce a technologically mediated ritual of visual piety. From the beginning, Phalke recognized in film the potential for the magical and the quasi–sacred extension of traditional icons. His express aim was the bringing forth of the Hindu pantheon through the 'technical magic of the cinema; to revive the gods of the *Puranas*; to fortify and gladden the masses of India in a moment of national self–affirmation' (Kapur 2000b). While the producers and directors of the era of mythologicals that ensued may have had a combination of commercial and ideological motivations, the viewing audience was quick to accept the darshanic and soteriological nature of this 'kinetic transformation of hitherto static iconography' (Kapur 2000c). Yves Thoraval (2000) notes that 'audiences were known to carry out prayers and rituals in the halls' to the silver screen manifestations of their adored deities.

Aathi Parasakthi

Here is a first person account of the cinema hall response to the screening of the 1971 Tamil devotional film *Aathi Parashakthi*, directed by K. S. Goalakrishnan. 'I was very, very young when I watched this movie . . . in the theatre. Many women in the audience would go into a trance when Padmini started singing 'Aayi Magamaayi', so much so the management had *viboothi* (holy ash) on standby, for this song!!!! I remember being terrified when it happened. Such was the power of the situation and music.'[7] At the heart of this sequence is a multi–step zoom to a close–up of the deity *Mariamman* in a cinematic enactment of *darshan*. This moment is preceded by a long shot of the Shamaness, played by screen 'idol' Padmini, dancing beyond the sanctum, shot as if seen from the point of view of the deity. This cuts to a long shot of the deity and then the zoom. This is followed by a tracking shot of the shamaness rushing forward towards the deity along the previously established line of vision, accentuating the spiritual or psychic identification of the seer and the seen. In the musical number that follows, Padmini ascends to a platform, strikes an iconographic pose and then proceeds to morph, by way of simple but effective filmic dissolves, through the significant forms of the goddess –Meenakshi, Visalakshi, Kamakshi, Mariamman. In this sequence, with its iconographic formality and prolonged frontality, the morphing goddess is undoubtedly presented to the viewing audience for *darshan*. Later, in the central mythological image of the film, the goddess, as Durga, slays the demon Mahisasura. As the tacky special effects battle reaches its climax, the protagonists arrive at the iconographic slaying image and hold the pose to abstract lighting effects –indexing the iconic import of image and presenting it for *darshan*. This highlighting, or extracting from the narrative flow, of the iconic image is in keeping with the modality of folk theatre. For instance, in the *Chhau* (masked dance) presentation of the *Mahishasura–Vadh* (slaying of demon Mahisha) from Purilia, when this same scene is reached in the performance, the tableaux is held and rotated to give *darshan* to the surrounding audience.[8]

Deivam

The renowned musical discourser Thirumuruga Kirupananda Variyar appears in person at the start of the film *Deivam*, whose contemporary melodrama is intercut with footage of a live *Kathaprasangam* or sacred musical discourse. The venerated discourser has also graced the films *Thunaivan* and *Tiruvarul* with his presence. Indeed the sanction and participation of this revered personage in such motion pictures is indicative of the institutional acknowledgement of the religiosity of popular *bakthi* cinema. This has been so from the early days of the mythologicals. For instance, the saint film *Sankaracharaya* (1939) was made with the blessings of the contemporary *Sankaracharaya* of Sringeri Mutt, who even presided over its premier (Baskaran 1996). In *Deivam* Kirupananda Variyar begins by proclaiming that Lord Muruga has, from *Puranic* times, been present and performing his miracles in the world. He states that in the onward rush of today's materialistic world, although the Lord's miracles still persist, they are hidden in the hearts of silent devotees. Then, in the words that open this paper, he draws an analogy between musical discourse, poetry, sculpture and cinema and endorses this new medium. As his words index the moment of narration, bringing it into a perpetual cinematic present, his gaze addresses the camera for a darshanic[9] close up. He then introduces the narrative –at Marutha Malai, a famous shrine to Lord Murugan at Coimbatore, during the *Ther* (chariot) festival, a notorious thief comes to steal the jewels that adorn the statute of the lord . . .

Bibliography

Baskaran, S. T. (1996). *The Eye of the Serpent*. Madras: East West Books, p. 14.

Burckhardt, T. (1976). *Sacred Art in East and West: Its Principles and Methods*. (Trans. Lord Northbourne). London: Perennial Books Ltd., p. 9.

Coomaraswamy, A. (1935). *The Transformation of Nature in Art*. Cambridge, MA: Harvard University Press, p. 26.

Coomaraswamy, A.K. (1977a). 'Editor's Preface', in Coomaraswamy, R. P. (ed.). *The Door in the Sky*. Princeton: Princeton University Press, pp. xii–xiii.

Coomaraswamy, A.K. (1977b). 'Mind and Myth', in Coomaraswamy, R. P. (ed.). *The Door in the Sky*. Princeton: Princeton University Press, p. 4.

Coomaraswamy, A. K. (1977c). 'On the Indian and Traditional Psychology, or Rather Pneumatology', in Lipsey, R. (ed.). *Coomaraswamy: 2 Selected Papers Metaphysics*. Princeton: Princeton University Press, pp. 333–78.

Dalmia, Y. (2001). *The Making of Modern Indian Art: The Progressives*. New York: Oxford University Press, p. 23.

Eck, D. L. (1998). *Darsan: Seeing the Divine Image in India*. New York: Columbia University Press.

Gardener, P. (1982). 'Creative Performance in South Indian Sculpture: An Ethnographic Approach'. *Art History: Journal of the Association of Art Historians* (Onias, J. ed.), vol. 5, no 4. December Issue. London: Routledge & Kegan Paul, p. 478.

Guenther, H. V. (1975). *Mind in Buddhist Psychology*. (Kawamura, L. S., translator). Berkeley: Dharma Publishing, p.. 1.

JVC Video Anthology of World Music, Tape 12: India 2, Smithsonian/Folkways Recordings, Victor Company of Japan Ltd.

Kapur, G. (2000). *When was Modernism: Essays on Contemporary Practice in India*. New Delhi: Tulika Books, a) pp. 237–8, b) p. 237, c) p. 249.

Lovejoy, M. and Jacob, P. (1999). 'Negotiating New Systems of Perception: Darshan, Diegesis and Beyond', in Ascott, R. (ed.). Reframing Consciousness. Bristol: Intellect Books, pp. 61–6.

Morgan, D. (1998)*Visual Piety: A History and Theory of Popular Religious Images*. Berkeley and Los Angeles: University of California Press, p. 1.

Prasad, M. M. (1998). *Ideology of the Hindi Film: A Historical Construction*. New Delhi: Oxford University Press, pp. 72–6.

Thoraval, Y. (2000). *The Cinemas of India (1896–2000)*. New Delhi: Macmillan, p. 6.

Notes

1. *Muruha Perumal, Arumuga Perumal, Skanda Perumal* and *Cenpaha Perumal* are all names of Lord Murugan, the Son of Shiva and the adored God of the Tamils.
2. Quoted from the Venerable Tirumuruga Kirupananda Variyar, who has been nominated as the 64th Nayanar or Saint of the Shivite tradition, in the opening sequence of Dandayuthapani Films' Deivam. Transcribed, translated and edited by the author and Sathiavathy *Deva Rajah*.
3. Pneumatology is the branch of Christian theology concerned with the Holy Ghost. More generally it is the theory of spiritual beings. It is used here in Ananda Coomaraswamy's (1977c) sense of a metaphysical, as opposed to an empirical, psychology.
4. Soteriology is a 19th century term that refers to the Christian doctrine of salvation. In today's usage, however, it is applied to index a broad range of doctrines of release from the mundane dimension existence.
5. India's first feature film, *Raja Harishchandra*, was made in 1913 by Dadasaheb Phalke.
6. Ananda Coomaraswamy disparaged these images as being theatrical conceptions, wanting in imagination and lacking in Indian feeling in the treatment of sacred subjects (Dalmia 2001).
7. Tamil Film Music Portal Forum, Thread: Hits of KVMahadevan, From: NOV(@ 02.184.134.10), Posted on: Friday 19 December 22:20:29 EST 1997. Last accessed: 22 June 2003. Available at: www.tfmpage.com/forum/13597.15976.13:10:44.html.
8. *JVC Video Anthology of World Music, Tape 12: India 2*.
9. Darshan is not restricted to icons but is also applicable to temples, sacred landscapes and holy personages.

5.6 Search For Utopia: Human Consciousness and Desire

Julia C. Rice

Imagine a world in which it is safe to walk the streets at night. Where crime has been eradicated and everyone cares for each other; a place where friendship can flourish and sickness is a thing of the past. In 1516 the author Thomas More wrote of such a place. He named it Utopia. (More 1998). To More, Utopia was a mythical island where everyone was equal in both their abilities and responsibilities yet, more importantly, they were content with their existence. Although Moore may have given a name to this place, it was not solely his vision. Many people dream of discovering such places yet they rarely seem to find them. Regardless of how close the realisation of a utopian existence appears it is often elusive, remaining just beyond reach.

Utopia, then, has become more of a state of consciousness than a physical place and is associated more with temporal and aesthetic, than with demographical, significance. Several people, however, have attempted to give their vision of Utopia a physical form. The architect Le Corbusier realised his visions of Utopia in his representations of planned theoretical cities in which skyscraper apartments were erected in order to provide a compact living environment while Walt Disney –the founder of the Disney Company –strove to create an environment in which he was surrounded by like–minded people who were subservient to his ideologies.

The word 'Disney' inspires confidence and is synonymous with childhood dreams. Yet to dream is to be passive and in touch with the unconscious mind, whilst the act of creation –thus giving dreams a physical presence –requires a state of heightened consciousness. While some are content to simply share Disney's visions of Utopia, others have also been driven to make their visions a physical reality.

For some this has also taken the form of creating films while to others it has been simply to identify with the protagonists in both their screen and public personas in an attempt to escape their mundane existence, even if only for a short while. The ever–increasing number of tribute bands and the success of television programmes like *Stars in Their Eyes* that encourage unknown vocalists to try to replicate both the appearance and the sound of well–known celebrities are a testament to this.

To the audience the lives of television, pop and film stars are to be envied and desired; they live in luxurious homes and wear only the best designer labels and perfumes. They are seemingly either loved or envied by everyone and live extravagant lifestyles. Despite the fact that this is not a true representation of the day–to–day lives of the stars, to their adoring fans they appear to live a utopian existence. The audiences' interaction with film and television has always been subject to considerable research. Just over 50 years ago educationalists were worried that 'television viewing would displace reading' (Gunter 1997). However, media reports that J.K. Rowling's most recent book –*Harry Potter and the Order of the Phoenix* –sold 1.8 million books in its first day of release would seem to dispel this belief. In fact –as with other forms of media –television has been the inspiration for numerous books, both on the medium and the programmes that it broadcasts. More recently television has been criticised because:

> *It is seen as having a deleterious influence on family interaction –family members no longer talk to each other as they used to –leading to a breakdown in the essential bonds that are so crucial to a stable family environment and to the development of socially responsible children.*
>
> (ibid)

This was brilliantly illustrated in the Ealing film *Meet Mr. Lucifer* (1953), in which the television was seen as an instrument of the devil, used to destroy people's ability to communicate with each other and therefore leaving them vulnerable to domination. In reality this particular film was more indicative of the studio's own fear of redundancy owing to the mass popularity of the 'new' medium rather than concern for society as a whole. However, as identified by John Ellis in his book *Visible Fictions*, the two media were never 'in direct competition with each other: broadcast TV cannot wipe out cinema any more than cinema was able to wipe out theatre' (Ellis 1992). Although film and television are both forms of audio–visual media they are received very differently.

Film is considered a primary media which has the ability to enhance learning owing to its ability to hold the attention (Gunter 1997). However, television is considered a secondary media as the viewer often watches it while involved in other tasks such as preparing meals, writing, etc. Despite this, the idea of television as a tool to suppress and control the viewer –possibly inspired by George Orwell's 1949 novel *1984* –has endured and not without good reason. Through the introduction of regulatory governing bodies –such as the British Board of Film Classifications and the Independent Television Commission –and encouraging self–censorship by the media, the government is able to restrict the type of information and entertainment that is available to the public. The editing of the material also substantially influences the way that it is received by the audience. However, it has emerged that one form of mass communication –once restricted to the realms of the privileged –is proving very difficult to control.

Owing to its global nature, and its growing popularity, the Internet is difficult to govern.

Despite this, the Internet and other advancing media technologies continue to be –though often with apprehension –accepted and welcomed in both the home and the classroom. Through this easily accessible media –which can be an excellent source of education and entertainment –users are no longer restricted to identifying with the actions of characters but can create new identities and personas for themselves. The anonymity of the medium is very appealing to those who lack confidence in public and provides the consciousness with an opportunity to escape the physical restrictions of the body that it inhabits. However, it is not the use of the Internet by adults that is proving problematic but that of children. Incidents where adults –possibly paedophiles –have masqueraded as children online to gain the confidence of children are often published in the press and the ensuing moral panic resulted in a £1 million government initiated advertising campaign to make parents more aware of the dangers of the Internet. According to Home Office Minister Hilary Benn:

> *The Internet has opened up a new world for children which is educational, informative and, most of all, fun. But we are aware of the potential for paedophiles to misuse modern technology to abuse the trust that children place in them by attempting to 'groom' them through chat rooms.*
>
> (wiseuptothenet 2003)

The fact that the Internet has the potential for misuse is not in question; however, the way in which children use the Internet and the effects that it has on their developing consciousness is. Like film, the Internet is considered a primary media as it has the ability to hold the attention. Unlike early television, the Internet is very interactive yet at the same time often requires physical isolation –something that has proved problematic with the recent convergence of the two forms of media by cable and satellite companies. Therefore the fear is that this form of media could prove detrimental to the developing consciousness of the young, rendering them unable to function in face–to–face social situations or to accept the restrictions placed on them by those in positions of authority. Initial primary research, which was carried out on several large groups of Welsh children aged between 11 and 15 years old from a wide variety of social backgrounds, would seem to dispel some of this fear. It identified that only a relatively small number of young people –less than 4 per cent of those questioned –preferred to use the Internet to make new friends rather than to do so in face–to–face situations such as in an educational environment.

This was more evident in those who claimed to have used the Internet from a very early age (under seven years old) than it was of those whose first experience of the Internet was around ten years of age. Although it could be argued that early use of the Internet is consistent with those children whose social skills would also seem to be underdeveloped, this is an area that will require further research. There is evidence to suggest that those whose social skills are underdeveloped are also weak academically; therefore other factors such as disability, social background, etc. have to be taken into consideration. One factor was conclusive in those who were questioned: the majority claimed that their intended use of the Internet was to converse with existing friends, therefore reinforcing their existing physical friendships, and to gain information on games, films and other interests including hobbies –which provided them with conversational material –rather than to seek out new ethereal friendships. They were aware of the fact that any information that they received from strangers was unlikely to be truthful as they were also giving out inaccurate information about themselves. The ability to do this –and to be able to converse in private without the fear of interference from adults –was cited as one of the main attractions of the Internet and was especially important to those who felt that their lack of friends was the result of their appearance or social background.

Around 20 per cent of those questioned felt that their use of the Internet at home had affected their social life saying that it restricted the number of people with whom they came into contact –owing to the fear of conversing with strangers in public and on the Internet –rather than broadened it. It was identified that the majority had only developed one group of friends –those that they made in school –rather than also socialising with different groups of children who live in their neighbourhood, thus creating a very closed world to exist within. Very few of those questioned felt that their parents believed that Internet use had affected their children's social life and considered it an excellent learning tool. From their responses it was evident that, despite its promise of an ethereal utopian existence, these particular children were aware that when dealing with multimedia products the disavowal of the problematic is counter–productive as technology is not a seamless product. When dealing with computer technology things often go wrong, therefore making it difficult for them to accept any association with a utopian existence.

Although the initial primary research has identified that these particular children were not using the Internet to create their own utopian world, its use would appear to be having an effect on their social development and therefore could be having long term effects on the development of their consciousness. It is thus essential that the research into this area continues over a period of years so that any effects can be carefully monitored and documented.

Bibliography

Ellis, John (1992). *Visible Fictions*. London: Routledge.

Fishman, R. (1977). *Urban Utopias in the Twentieth Century: Ebenezer Howard, Frank Lloyd Wright, Le Corbusier*. New York: Basic Books.

Gunter, B, and McAleer, J. (1997). *Children and Television*. 2nd ed. London: Routledge.

More, Thomas (1998). *Utopia*. London: Orion Publishing Group.

Wiseuptothenet www.wiseuptothenet.co.uk/ho_model.pdf

http://213.121.214.245/n_story.asp?item_id=325 Accessed 6 January 2003.

5.7 Super Interactivity: Art, Consciousness and the Dawn of the Participatory Age

Alex Shalom Kohav

When the medium was participation

The history of art and the history of consciousness coincide in unique ways. Both can be seen as divided into three distinct, immense periods:

1) Participatory Eon (ca. 20,000 BCE till ca. 500 BCE

2) Observational Eon (ca. 500 BCE till ca. 2,000 CE)

3) Participatory Eon (ca. 2,000 CE, and on)

In the first Participatory Age human beings were fully immersed in a magical participatory process that determined the course of their lives. No one at that time would have called his or her activities art, as that concept belongs to the subsequent era. Nevertheless, what we might call today 'art' shaped people's 'consciousness' to an overwhelming, almost unimaginable degree.

What was it that human beings participated in? What was the first 'art' on earth? Scholars would give to their activities and/or state of consciousness a designation 'participation mystique' –a curious term that arguably has obscured more than it revealed.

It is clear, however, that in the first Participatory Era people were in touch with much wider dimensions of reality than we are today; and through their participation in ritual 'art–making' activities, they were becoming co–partners with manifold forces and entities that in turn shaped their consciousnesses and thus their destinies. Their 'Medium' was participation and the 'message' was, 'yo are in charge'.

The magical First Eon stretched from the original womb–like enclosure of the prehistoric cave, some 20,000 years BCE, and encompassed the sophisticated Egyptian civilization that centred on another womb–like enclosure, the tomb. Finally at about 500 BCE, the bright sunshine of the Greek logical consciousness all but destroyed the magic. The Observational Age began.

Greek sensibility demanded logic, clarity, visuality –in short, left–brain activity. The visual was and remains so much the essence of the Greek approach which our western mind has internalized that 'art' can hardly be conceived by us as being other than 'visual art'.

Yet what our semi–wild ancestors were after was not at all visuality. Their aim was to affect their consciousness in a manner that would bring about the desired energy for survival, perhaps even for flourishing. The visual was a part of this but not the only or even the most dominant part. Moreover, as physical visuality was not overplayed while allowing other methods to be equally active, the conditions were perfect for the magical effect of psychic visuality, within the right–sided brain.

There was another road available, an alternative to the Greek 'down–to–earth' converging on the left–brain–centred lower consciousness. That other, optional road –instead of the Greek 'observational' mode –was the original Hebrew Kabbalistic approach of Moses, a participatory road to an expanded, or divine, Mega–Consciousness.1

Such a consciousness is not simply finding its home in the right half of the brain only but is nourished through no less than ten spiritual sensing points situated along the human body (including also the left side of brain). These are the little–known Ten Sephirot of Kabbalah –which are the elusive 'image' that we humans share with the God of the Hebrews.

The birth of window vision

What followed that magical first Participatory Eon, in terms of human consciousness and art, was the Observational Age. People no longer participated in any ritual–cum–art that once was meant to affect their consciousness and therefore their destinies. Art was now meant to be looked at, observed, marvelled at or disapproved at will but no longer to be participated in.

Such a radical shift in human consciousness brought about drastic changes and innovations in perception. The most significant change was the invention of the 'window' vision. (Thus it is rather disingenuous of Bill Gates to claim credit for something that was introduced two and a half millennia ago!)

Now people viewed the world as though through a window. The world suddenly was outside of themselves, while their 'selves' and their private lives were on the inside. The artificial, man–made chasm between 'body' and 'soul' was born. The 'window' frame of vision became the bridge that connected these two separate realms which earlier, in the magical Participatory Eon, were one and the same.

Art faithfully followed the new sensibility and promptly adopted the 'gaze through the window' approach. The very idea of 'art' was conceived right then and there, when this joint product of the Muses and inspired craftsmen was being torn away from earlier magic rituals. Since the result has been a rather unnatural –some even will insist fake –activity, the notion of 'art–ificial' was born simultaneously with it.

Thus the Observational Age separated consciousness from 'art'. The primacy of seeing and looking, rather than experiencing, was established from then on. The motto of that eon, in art, was: ~ The meddium is the message ~ while in consciousness it is, 'seeing is believing' and 'a picture is worth a thousand words'. The once shamanic artist became an inspired craftsman at first, the bohemian clown later.

It was only a matter of time before 'gallery' appeared on the scene. What is a gallery, essentially? 'Gallery' is the place to look at 'art', as everyone knows. 'Look but don't touch', is the byword of the artworld which is therefore more restrictive than the general retail industry, which at least allows one to handle its wares.

Of course, the gallery is also a place to buy art, not just to gawk at it. By turning art into a commodity to be purchased, the separation between 'art' and 'consciousness' was completed. Something purchased for purely sensuous pleasure (or investment) cannot transform your life or affect your consciousness, other than perhaps titillating it and, in some cases, polluting it.

Crisis of the relevancy of art
An extreme crisis in art has been reached –the crisis of the relevancy of art. A further detrimental aspect has appeared in the form of psychic pollution, as art is forced to compete with the cinema, the TV and advertising and to utilize shock and/or entertaining tactics.

Has anyone conducted research into the impact of all this activity on the consciousness of exhibit visitors (let alone of those who do not visit)? The impact is clearly beneficial, in terms of being an economic stimulus for the areas that encourage it.

The effect on viewers' consciousness, however, is not simply miniscule. A case can easily be made that today's art is as often detrimental as it might also be simultaneously entertaining. Entertainment, however, is a tool for easing the passing of time, not for enabling transformation of consciousness.

In an effort to attract visitors away from a myriad of powerful amusement options, ambitious institutions tend to promote art that is either highly entertaining or shocking in some way. The result is invariably the 'pollution' of the psychic environment that is already quite toxic as it is.

Remember the shock of Julian Schnabel's broken ceramic plates attached to the canvas? This was a mere couple of decades ago. Since then we have moved light years away from plates, umbrellas, chairs or what have you, glued or otherwise affixed to the –by now turned pathetic –canvas.

The idea of the 'picture' itself had to go altogether, as we needed something to better match our confused, disjointed post–modern lives. Something more akin to Damien Hirst's cow parts floating in formaldehyde or Matthew Barney's curious creatures crawling on video screens. These, however, merely added to the overall psychic pollution.

We have arrived at the last, crumbling frontiers of observational shock–and–entertainment art. For such art, the medium is still the message. Even in the current Internet–based explorations that force upon the mostly sedentary population a new art form, to partake in while seated.

Participatory (Super Interactive) art: new caves for our consciousness

The demise of the Observational Era is upon us. Humanity has recently entered a new Participatory Age once again. New winds are blowing, disrupting the obsolete modus vivendi of the Observational Eon's art and its associated 'polluted' consciousness.

Such 'observational' works as Cezanne's peculiar green mountain vistas and Pollock's Number One –not to mention the ancient Chinese contribution –will, of course, be always gratefully acknowledged. But now mountainous vistas and an awareness of a One all–pervasive force must be developed from within, from the engaged public's expanding consciousness. This can only be accomplished through a participatory art, an art that is not necessarily visually sumptuous but which can give birth to inner visuality and provide an opportunity for transformation of consciousness.

In the first, archaic Participatory Eon, enclosures such as caves and tombs imitated that primal enclosure which nurtured us all into life –the womb. Thus consciousness and art engendered by such total environments tended to be organic, Gestalt–like, nurturing and life–sustaining.

Today we discover a similar tendency reborn, at a higher turn of the evolutionary spiral. New spirit–nurturing, consciousness–transforming 'caves', 'tombs', and 'wombs' are being created.

It may have all started with the advent of installation art, which moved away from simple two–and three–dimensional art forms that were meant to be viewed by a viewer. Installation art forced the viewer into a complete environment. The visitor, though still a viewer, was now separated from his/her habitual condition and circumstances. The potential for transformation of consciousness had been born.

The new Participatory Age abandons the observational mode in art and consciousness alike and adopts a participatory one in its place. The new caves turn the visitor from a viewer into a participant.

An illustration of this is a cave, or CAVE, that has been developed by the University of Illinois at Chicago since the 1990s. CAVE stands for 'CAVE Automatic Virtual Environment' and consists of enveloping computer monitors and VR gear. It is a 'networked virtual reality . . . VR worlds that distantly located people can view and change in real time'. as their website explains.2

There are many other examples today of similar silicon frontiers that are enabling a dramatic expansion of human options.

My own explorations of Kabbalistic methods of attaining altered states of consciousness convinced me that there is nothing more important for human beings than expansion and transformation of consciousness.

Why should art, which ostensibly is about highest human aspirations, be instead bogged down by the aspirations of a consumer society, complete with such by–products as psychic and/or visual pollution? Could art be, instead, a vehicle for the transformation of consciousness, perhaps similar to the ways of the first Participatory Eon?

To this end, I began to develop 'Super Interactive,' or 'participatory', installations. In these, the visitor no longer observes what has been displayed for him/her to look at. Rather, visitors engage in activities that have the potential to transform consciousness, sometimes on a profound level.

An example of such an installation is Senselessness, in which visitors can experience the so–called Floatation Tank (a.k.a. Isolation Chamber). The tank enables the participant, probably for the first time in his or her life, to experience 'sensory deprivation'. Similar to the virtual CAVE at the University of Illinois, it is an example of a 21st century 'cave', in the sense that it decisively separates the participant from his/her normal environment.

Other examples of my Super Interactive projects from 2001 to 2002 include such installations as Dali Llama (participant occupies a cage next to a caged live llama animal); Balls (playing ping–pong in an environment saturated by balls, real and electronic); Gravitation Dissolving (experience an approximation of weightlessness, via a Virtual Reality hang–glider simulator); Bullet–Proof Body

Condoms (participants try on huge plastic body bags, in a simulated effort to protect themselves from chemical and biological terror); The Wizard of Schmooze (a 'disembodied', invisible 'wizard' interacts with participants); or Art Should Touch You (visitors break the taboo of not only touching the artwork but even of being themselves touched by it).

These projects liberated me from the tyranny of the observational approach. Subsequent work involved more specific and direct ways to both imitate the prehistoric cave and the primal womb, as well as to utilise latest technological means for beneficially effecting participants' consciousness. I utilise the principles discovered from my general explorations of human consciousness, especially through personal efforts at reconstructing the most potent –yet practically forgotten –original Kabbalah of Moses.

My newest installation projects (2003) include The Womb (where participants swim in a tank of water, being showered by coloured lights and custom sounds); Cube of Existence (experiencing five–dimensional Hyper Cube which human beings inhabit and sense as 'reality' or 'existence'); and such series of projects involving enclosures as: Primal Hollow–grams; Cells; Chambers; Caves; Containers; Coops; Cocoons; Enclosures; and Vessels.

There is no 'window' vision in such art anymore. Nor would the earlier motto of the Observational Era –'The medium is the message' –apply here. The motto for the new Participatory Age is: There is no message, only an opportunity for viewer transformation.

Mega consciousness and the coming 'doors' sensibility
Instead of a window vision, advances in technology will enable artists to reflect a new 'door' sensibility in their consciousness –and art. These will be virtual doors, through which a participant will be able to enter diverse distant or formerly out–of–bounds worlds.

Through a marriage of art and technology, human beings will be dramatically influencing their consciousness and thus their lives. Anything will become possible and viewer transformation via art will become the state of the art, the norm as it once was, long ago. Art once again can become a powerful tool for humanity's transformation of consciousness.

At that point, if not before, human beings will begin to differentiate between various states of mind or levels of consciousness. Today this is still a dream, and in our still nebulous understanding of consciousness, such concepts as Mega Consciousness are banished to our intellectual fringes.

World as immense interactive installation

Aesthetic sensibility is not only intoxicating through its approach that engages our senses and imagination. It can also be a crucial revelation tool.

The world, for example, can be seen as one vast Super Interactive art installation, with each one of us being both visitors/viewers and participants. But aesthetics would have little to occupy itself and would resemble the semi–dead or anaesthetized art criticism of today if its object of inquiry, the arts, did not have moral issues to scrutinise, as seems to be the case in our sanitised present. The dearth of moral concerns today by no means testifies to our society's high morality, only to us being severely tranquillised.

As has been the case throughout the history of human civilisation, art and consciousness go hand in hand. Our liberation must start with our own consciousness and art must not simply reflect this process in some kind of neo–representation but itself be one of the direct conduits and means of liberation.

Notes

1. Although Rabbinical Judaism lost its connection to that original Hebraic road, it has begun to experience a rediscovery and revival. This writer has worked at restoring the Kabbalistic system of Moses and is completing a book on the findings.
2. EVL: Alive on the Grid www.evl.uic.edu (as of September 2002).

5.8 Pete and Repeat were Sitting on a Fence: Iteration, Interactive Cognition and an Interactive Design Method

Ron Wakkary

The important thing is to make a start.

(Umberto Eco 1997)

Introduction

This paper is about the exploration of an alternative design method. The exploration arises out of the awareness of the need to invest more design energy and creativity in the area of 'framing the problem' and the need to design 'in–the–world' throughout the design process. The exploration and discussion is based on a design method concept of interactivity. This concept of interactivity, not to be confused with interactive design, i.e. web and interface design, is based on Henrik Gedenryd's notion of 'interactive cognition' and Donald Schön's ideas as embodied in his concept of 'reflective practice'. The paper discusses the related elements from Gedenryd. It describes preliminary principles for an interactive design method and presents examples from a research project on an augmented reality interface for museums, known as ec(h)o.

Iteration

Iteration is a process common to many design methods and it is often valued as critical if not quintessentially a part of 'good design' practice. Iteration in design methods is almost universally seen as an exemplar of flexibility in current design. If a designer wants to point out the innovative or robust nature of his design process, he will quickly demonstrate its iterative qualities, that is, its ability to repeat. As the HCI (Human Computer Interaction) computer scientist Bill Buxton aptly states it, 'keep trying until you get it right' (Buxton and Sniderman 1980 p. 72). To design iteratively means that the designer can repeat at each step in the design process, go backwards or the designer can repeat the entire process again and again. The ability to go back to the beginning of a phase of a design or the previous phase if something 'goes wrong' is seen to be the ultimate in flexibility and demonstration of the evolving corrective response mechanism within the design process. Another form of iterative design is the ability to 'rush' through the entire sequence of the design process in order to get user feedback and to emulate designing 'in–the–world'. The idea is to present the outcome to end–users as a prototype, thus facilitating the collection of data from end–users, so that the designer can repeat the process again and develop another prototype and collect more data, repeating the process until the designer 'gets it right'.

However, if we scrutinize the concept of iteration closer, we find that not only is it not an exemplar component of current design methods but rather it is part of the undoing of the sequenced design method. Further, it is symptomatic of all that is wrong with most design methods today. Iteration is what the designer/cognitive scientist Henrik Gedenryd refers to as an 'ad hoc extension' employed to overcome the weak and unrealistic aspects of most design methods (Gedenryd 1998 p. 97). For example, the need to repeat phases of the process such as 'analysis' or 'design' phases until 'one gets it right' may be more symptomatic of the typical lack of attention and energy put into specifying the problem. Framing the problem is absolutely key to the entire design process yet typically it is not explored or, worse yet, it is determined by others and adopted by designers as a given. The other form of iterative design, the 'cycling' through design process many times in order to 'iterate' prototypes and to have those prototypes come in contact with the 'real–users', is an attempt to design in a 'real–world' context. However, it begs the question of why not 'bring in' users, or more simply, people, into the design process earlier? Why design a process where the real context is not tested until the end and therefore rush through the process so you'll get to the end sooner?

The model of technical rationality

Henrik Gedenryd's 'How Designers Work' (1998) analyses the design process as validation of his theory of interactive cognition. While the aim of his thesis is cognitive science he found strong commonalities with design, such that his thesis comprised a joint critique of design and cognition. Gedenryd's central argument is that the traditional models of cognition are based on a 'pure mental model', what he refers to as 'intramental', whereas evident in design practice, cognition is an interactive process that acts and responds to the world that it is in, it integrates and combines cognitive activities with events in the world in order to act and reflect 'in the world'. He argues that a truer model of cognition, what he refers to as 'interactive cognition', lies within the genuine practice of designers.

The critique is therefore aimed at the formal levels of the two disciplines. It assumes, quite rightly, that a large gap exists between reality and the formal descriptions. More importantly, the formal descriptions share an underlying pattern that is a general model of rationality and rational action (ibid p. 55). Gedenryd finds the rational model pattern by analysis of a wide range of formalizations, including classical design methods, software engineering, folk psychology, cognitive planning theory, problem–solving theory and information–processing theory. Four principles emerge from the analysis. The principles are shared by traditional design methods and cognitive models (ibid p. 115):

1) Separation: the separation of the design process into distinct and isolated phases; in cognition, the separation of thought and action

2) Logical order: in design , the explicit order or sequence of activities; in cognition, thought precedes and determines action

3) Planning: the pre–specifying of an order in which to perform the activities within a phase; in cognition, plans as the mechanism that predetermines action i n thought

4) Product–process symmetry: the structure of the process of design or thought is reflected in the outcome or action

Gedenryd's analysis and demonstration of the roles of these principles is more detailed but for our purposes it is ample evidence to demonstrate the principles within the most generic models for design methods. Together, the principles of 'separation' and 'logical order' generate the basic three–stage design model of analysis, synthesis and evaluation. That is, each step is distinct and takes place in isolation and follows a logical sequence. The more elaborate design model demonstrates the interdependencies between the phases or how strategies within one phase plan and structure the activities in the subsequent phase. For example, Gedenryd points out, a prototypical example is the final part of analysis plans the course of action for the synthesis or refinement stage before this activity even begins.

Not surprisingly, Gedenryd's critique of design methods led him to consider formal design methods as complete failures. Behind these failures was the false move of basing design methods in the model of rational action, rather than the actuality of practice. While the concept of design as a science has not gone uncriticized, it is persistent at the foundational level and continues to hold sway, particularly in the area of design methods.

Iteration again
Returning to the idea of iteration, we may ask how does iteration contribute to design methods' success or failure. In Gedenryd's brief analysis, iteration is an idea that counters the model it is intended to support:

> ...*iteration is a prototypical ad hoc extension, that is, an ill–considered added feature that handles a certain condition, but which in doing so goes against the original idea, and is therefore incompatible with it –thereby, in reality it constitutes no solution at all.*
>
> (ibid p. 97)

We could say that iteration is a roundabout confirmation of Gedenryd's critique of the rational model. Gedenryd elaborates on how iteration disassembles the principles of 'sequence' and 'logical order':

> *By allowing for iteration, a stage model comes to saying that you can do anything, in any order, as many or as few times as you like. By allowing for everything, it no longer says anything about their order. But if you do that, you have given up what was the purpose of these models in the first place: to specify what things to do, when to do them, and in what order, so as to guide the designer. The only substance that remains is a list of the activities that are included.*
>
> (ibid pp. 97–8)

Iteration is a workaround that illustrates that the idea of separated and ordered activities do not hold. Iteration utilized to the extreme reduces the typical design method to no more than a list. Design methods become a checklist of activities that can be completed in any order, sequence and frequency. For Gedenryd, this confirms a central understanding of interactive cognition and design: 'design consists of several component functions that cannot be held apart, and that display no general ordering principle among them' (ibid p. 98). In addition, the earlier discussion on iteration pointed out how iteration is symptomatic of two needs in design practice: the need to invest more design energy in 'framing the problem' and the need to design in the context of 'in–the–world' throughout the design process.

Interactive cognition

Key aspects arise out of the analysis of iteration in design. Design consists of several component functions that cannot be separated and have no general ordering principle. Design requires a great deal of attention and creative energy in 'framing the problem' and the design process acts best within an 'in–the–world' context. These aspects equally arise from Gedenryd's concept of interactive cognition. Gedenryd begins with the 'extended ontology of cognition' that moves beyond the classic model of 'mind and cognition' to include not only the mind but also action and the physical world. He advocates a shift away from the emphasis on the extension of cognition into entities –mind, body, world and actions –to activities or the interactions between entities (ibid p.12). Gedenryd's understanding of activity is rooted in the pragmatic inquiry of John Dewey and Schön.

A formalization of the theory in four steps is included below in order to provide a brief explanation of the concept of interactive cognition. Each step can be seen as a layer, which is made possible by the layers before it, capitalizing on them to successively add further advantages of the interactive mode (ibid p.115):

1) the rediscovery of the world: the advantages of dealing directly with the world instead of a surrogate; for example, 'position fixing' in ship navigation

2) manipulating the world –doing for the sake of knowing: the advantages added by action and interaction with the world; for example, the game 'Tetris'

3) fine grained interactive structure: maximizing the benefits of involving world and action; for example, a conversation: 'no to your right, no over by the quad, right there yah right there'

4) pragmatism enables specificity and shortcuts: a set of shortcuts made possible by drawing on the specific conditions of a situation rather than the general information a surrogate can only provide; for example, a conversation:
   a: Do you read?
   b: Do I read?
   a: Do you read books?

As an overarching example, Gedenryd sees in the process of sketching in design the incorporation of all four steps of interactive cognition.

Interactive design method

In combination with emergent actions in practice and a preliminary framework drawn from Henrik Gedenryd's notion of 'interactive cognition' and Donald Schön's 'reflective practice' (not discussed in this paper), the author explored the development of an interactive design method. Principles of the method include:

- design process is led by 'frame experiments'

- design components cannot be separated and have no general ordering principle among them

- design process is done 'in–the–world'

The term 'frame experiment' is from Schön. It is the framing of a problematic situation in order to make it more manageable: 'When he finds himself stuck in a problematic situation which he cannot readily convert to a manageable problem, he may construct a new way of setting the problem –a new frame which, in what I shall call a 'frame experiment', he tries to impose on the situation' (Schön 1983 p. 63). The techniques within the method include a 'frame experiment' in the form of scenarios and interactive workshops. The design process is led by 'frame experiments' that take the form of a scenario. Like traditional use of scenarios in design, the goal is to envision a possible outcome or future as a response to the design situation. The different forms of scenarios include role–playing, storyboarding, scripts/narratives, sketches, videos and interactive works. Although the process begins with a 'frame experiment' or scenario, subsequent scenarios are created whenever it is required in order to address the recurring phenomena within the design situation of 'complexity', 'uncertainty', 'instability', 'uniqueness' and/or 'value conflict' (ibid p. 39).

The recurring phenomenon of the design situation is enacted in interactive workshops as a way of generating design responses. Key features of the workshop include:

- workshops are 'planned one at a time'

- following the workshop is an 'interactive response' to the previous workshop, it arises out of the inquiry of the previous workshop

- workshops often are a response to frame experiments

- workshops can create the need for a frame experiment

In addition, the workshops strive to adhere to the principle of designing 'in–the–world'. To that end, people are part of the design process from the beginning. Workshops include potential end–users in a participatory design approach. The role of the designer is to design the workshop such that it frames for enactment. In order to invite participation, the workshops adopt a low–resolution approach, such as a paper prototype or the inclusion of prototyping as a participatory act within the workshop.

The method is very simple, involving the two techniques of scenarios and interactive workshops, yet the resulting process is a complex non–linear structure that does not include separation of activities or inherent sequences. It does not privilege planning preceding action or decomposition and analysis as a prerequisite for synthesis. It is non–linear and more web–like in its structure. The exact sequence is not predictable but the overall pattern is. For example, while you may not know what workshop will follow it is clear that it will connect with 'parallel' workshops and eventually a framing scenario. The workshops are structured to invite interaction and participation; this may be perceived as openness, yet at the same time, workshops are connected to scenarios that are quite closed.

Conclusion

In conclusion, the interactive design method replaces the linear sequence of separate activities with an interactive process. The result is a complex web of viable choices based on interactive engagement of designing in the world. Gedenryd's concept of interactive cognition demonstrates the connection between cognition and the act of designing. A natural outcome of his work is to reintroduce it as a conceptual tool for better understanding the design process.

Bibliography

Buxton, W.A.S. and Sniderman, R. (1980) 'Iteration in the Design of the Human–Computer Interface'. *Proceedings of the 13th Annual Meeting of the Human Factors Association of Canada*, pp. 72–81.

Gedenryd, Henrik (1998). 'How Designers Work: Making Sense of Authentic Cognitive Activities'. *Lund University Cognitive Studies*, no. 75.

Schön, D. (1983). *The Reflective Practitioner: How Professionals Think in Action*. New York: Basic Books.

5.9 Visual Art as an Earning Process: The New Economics of Art

Nicholas Tresilian

The world as it was
In days gone by visual artists produced closed easel–paintings in frames and free–standing sculptures on plinths –closed attractors (Tresilian 2000. pp. 171–5), tradable in an international marketplace. If all went well with an artist's career, the price of these products rose across the years. So long as the artist shrewdly husbanded a stock of earlier works and the current work was marketed by a dealer as interested in the living artist as in the dead, in course of time the value of the opus would grow to fund a handsome pension plan for a distinguished old age. It was a system agreeable to successful artists, private collectors and public patrons alike. It provided the private collector with artefacts in the hand, offering either favourable exposure to speculative risk or a hedge against fluctuations in the broad economy, plus a quick release of financial value on exit. It gave the public patron a reliable source of the high–value collectibles and protectables which are needed to justify the existence of museums and art galleries per se, enabling these institutions to act simultaneously as centres of scholarship and mass recreation, conferring high prestige on their native city or country, while helping it rake in the much–needed tourist dollar. As for the successful artist –in this sense no different from the rest of us –as the song says, there is nothing like a dime.

The world as it is today
In the last 30 years, however, the visual art of the western world has undergone a global sea change. Closed attractors such as paintings and sculptures, prints, drawings, etc. have lost their ancient monopoly of artistic value and are now openly challenged by a new genre of art based on the novel concept of the open attractor. The indoor installations and outdoor exstallations1 of the new 'large–format' art of the post–modern age systematically invert the value system of our traditional western 'small–format' art (see Table 1)

The modernist origins of large–format event–based art may be traced to the war–torn decade 1910–20: to the cafe concerts of the Dadaists, Tatlin's tower, Schwitter's Merzbilden and Duchamp's readymades. In the decades of global reaction against modernism that followed World War I, these initiatives were abandoned and lay largely forgotten. Decades later they were rediscovered, now as the archaeological foundations of the newly emergent large–format art of the later 20th century. If today small–format art represents the well–known past, then large–format art, with its wrap–around/walk–thru' imagery, seems to speak much

Small–format art	Large–format art
compact attractor	extended attractor
painting/sculpture	installation/exstallation
closed object	open event
'space'	'time'
eye outside looking in	eye inside looking out
contemplative	interactive
aesthetic value	proto–ritualistic value
d Nude reclining	d Nude descending a staircase

Table 1

more eloquently of our species' unknown but self–evidently risk–laden future (Beck 1992). There on the edge of chaos, between the organised past and the disorganised future, between the closed and the open attractor, is the dialectical epicentre of today's visual art.

Large–format art in the art–historical big picture

In a wider art–historical perspective, today's installations and exstallations may be seen as distant precursors of a new cycle of ritualistic art for a globalised economy in which our concepts of 'energy' and 'information' have become seamlessly –or to use Roy Ascott's term, 'moistly' –merged. In this sense today's large–format art has less affiliation generically with painting/sculpture per se than with the cave art of the hunters–gatherers and the temple art of the agrarians. But whereas Lascaux/Font–de–Gaume/Altamira and Stonehenge/Teotihuacan/Angkor Wat represent the majestic maturity of their respective cycles of creative development, today's large–format art in its travelling circus of curated annual, biennial, triennial and quinquennial exhibitions, is right at the opposite end of its own creative cycle –an emergent 'quantum foam' of intensely ephemeral images –art with a butterfly existence.

The transfer of creative attention to the large–format, open attractor may ultimately draw art into new kinds of productive engagement with the human economy (an engagement to which many 20th century artists aspired but were unable to achieve through the limited agency of the closed attractor). Alternatively it may come to populate the earth with creative ivory towers, imaginary Shangri–Las and virtual Glass Bead–Games. We cannot second–guess the future and would be wise to recognise the existence of a Doppler effect in all cultural evolution –such that events and images receding into the past have a different resonance for us from events and images emerging into the present.[2]

The problem of ephemerality

For now, however, the intentional ephemerality of so much present–day art poses problems for the private collector and the public patron alike. For the private collector large–format images are difficult to buy, generally too big for domestic display and, being perishable, fail as a store of financial value. For the public patron with a larger but still finite space to display and store a collection of works of art, installations/exstallations are like so many cuckoos in the nest: eating up storage space, difficult to conserve, problematic to reconstruct, vastly expensive to send out on tour –all of which makes it excessively difficult to determine their legitimate financial value. (In these circumstances it is easy to understand the striking commercial success of the `Britart' formula for a mezzanine form of contemporary art: images scaled–up to the dimensions of large–format art and with some appearance of 'openness' –we can walk between the cut–up sections of Hirst's Friesian bull, for instance –but which in reality are constructed in the form of closed objects and can therefore be bought, owned and displayed in a conventional way –as in the Saatchi collection.)

For the artist, too, if motivated to produce works with an essentially ephemeral character, the problems of economic survival are arguably even more acute than for the traditional painter/sculptor. The installationist all too easily finds him/herself on a treadmill of endless form–filling, constant chasing after grants and fees, earning a fitful living at the discretion of an ever–changing patchwork of funding–body officials in salaried and pensionable posts. Furthermore, the artist –if determined to practise as an artist fulltime –may still be on the same unpensioned treadmill in 10, 20, 30 years' time –the only available alternative way of earning a living, a job in an art school and the struggle with the weariness of the weekend art–worker through the remaining most productive years of life.

Addressing the problem

Artists who are dedicated producers of essentially ephemeral works –the bubbles in the quantum foam referred to above –may be able to improve their financial situation by taking a more objective view of the potential product–streams into which their own work stratifies.

a) individual artworks –installations/exstallations of limited duration

b) commercial spin–offs from the exhibited artworks –in the form of prints, drawings, photographs, video images, etc.

c) the 'workings' behind each individual artwork –the letters, statements, sketches, specifications, applications, legal instruments, etc. which constitute the residual archive material

All three tiers of the work can be used to provide funds for the artist. We have considered 'a' already –the ephemeral artwork funded by fees. Let us now consider 'b' and 'c'.

Commercial Spin–Offs

The world's most brilliant exponent of the commercial spin–off is surely the Bulgarian artist Christo, the construction of whose spectacularly beautiful but always short–lived works –pink plastic seas, wrapped Reichstags and so on –is completely funded by the sale of the associated working drawings and the artist's prints derived from them. These more durable by–products of the main creative process afford the artist a back–door entry into the legendary liquidity of the conventional art–market –small–format works thus funding the large–format enterprise –while for the collector constituting an authentic 'piece of the action'.

But there are problems for the artist here too. Does the small–format product–stream undermine the statement about art made by the large–format enterprise? Where in the event is the true centre of gravity of the artist's total lifework? What are its unique brand values and how are they divided between the physical and the ephemeral opus? To a lesser extent these are problems for the collector also. Is the bought item 'art' or is it 'archive'? Will its value fade as memories fade of the original ephemeral event? If large–format art continues on its march, will small–format works relating to it end up 'beached' on the wrong side of an art historical and financial divide?

'Don't throw it away . . .'

Given due care in its assembly and preservation, the archive of the artist's sequence of project workings now also has potential financial value. It has long been the practice of museums to acquire the archives of distinguished musicians and writers. With the trend to the short–term/large–format image in visual art, it is inevitable that art museums will go down the same path. Indeed the Tate Gallery has only recently opened the generously–endowed Hyman Kreitman Research Centre which, along with 120,000 exhibition catalogues and over 50,000 books, already contains some 700 individual artist collections: personal papers, correspondence, diaries, sketchbooks, maquettes, artworks, photographs, posters and audio–visual material –and now has room for many more. As physical artworks of international standing become more difficult to acquire, public patrons such as Tate, in order to continue to grow their collections, will have no option but to spend more of their purchasing budgets 'downstairs' rather than 'upstairs' –on archive acquisitions. The personal archive thereby becomes the artist's nest egg.

To hatch this particular chicken the artist needs to be aware of the archival potential of the studio audit trail from the earliest moment –sorting, storing and keeping safe these keys to the kingdom of his/her career. And today this may not

Mode 1	Mode 2
mono–disciplinary individualistic Exclusive context–insensitive Birunaccountable	trans–disciplinary, Combinatory Inclusive context–sensitive Accountable

only be a preoccupation of the individual artist. If, to adapt some current jargon from science, the individual artist represents 'Mode 1 culture' in the table below, the new large–format art points more in the direction of 'Mode 2 culture'.[3]

In turn, mention of Mode 2 culture brings us to the subject of the SOAN.

Introducing the SOAN

Postmodernism has exposed the limits of individual enterprise in visual art. With the emergence of large–format attractors as a new platform for creative expression, art begins to change from a cottage industry into a service industry broadly responsive to issues of human survival on a crowded planet where 95 per cent of all known species are already extinct. To meet these new challenges, artists have begun to evolve a new genre of art institution, the self–organising artist network or SOAN. Artstation, Art–Language, Artist Placement Group (now Organisation and Imagination Ltd.) are recent UK variant examples of the SOAN, whose ancestry can again be traced back to the combinatory networks of the early Modernists in Dada, Russian Constructivism and the Weimar Bauhaus. A well–organised SOAN has the big–project capability that the individual artist generally lacks. It can more easily measure up to the logistical demands of the large–format attractor. And unlike the individual artist, accountable only to his/her own creative imagination (and in this respect very like the traditional Mode 1 scientist accountable only for pure research), SOANs have the institutional robustness to be accountable for projects undertaken in the public space –the agora. Thus, to cite two organisations this author personally knows well:

- the Artist Placement Group of London inserted artists in a catalytic role into national and local government organisations of its day: DHSS, Scottish Office, DES, Peterlee New Town, Birmingham Small Heath, etc.

- more recently, Artstation of Cardiff, Wales has used a combination of public space installation and video to invoke the lost identities of asylum–seekers within the European agora

However, it is also true –and again speaking from personal experience –that a large

SOAN may embody all the political problems of the EU in miniature. In this sense today's SOANs stand at the foot of a steep learning curve both creatively and organisationally. But as more SOANs propagate worldwide to meet the demands of the new large-format art, it seems likely that they will become the main drivers of a new market in archive material encompassing substantial records by many hands (and with them a raft of potential new intellectual property issues). It is significant that Tate Britain has just offered a substantial sum for the archives of the Artist Placement Group.

A1, A2..........An

Nothing is ever altogether simple in the relationship between art and economics. It makes sense for individual artists and SOANs to seek to derive additional income from the formerly hidden aspects of their working process. At the same time, if individual artist and SOAN archives are considered as the first two elements in an expanding archival series going forward, it is already possible to imagine an 'nth' element in the series in which the 'archive' might contain the whole of human history –to be balanced off against the possible threat of massive human extinctions in the present century and beyond (Rees 2003). In this sense every first and second level archive is a subset of the 'nth' level archive and has potential as a diagnostic resource for the human condition –a snapshot of the lifework of a member or members of a species perhaps only a few steps away from its evolutionary terminus. These considerations may offer a corrective to the hubris of the archive compiler. More importantly, they may come to furnish a new system of 4–dimensional perspectives for art itself. Large–format art may just now seem most obviously to offer new ways of interrogating our present world and its troubled future. But to be fully meaningful, at some point it will have to turn and take on our past as well –in the process, collapsing the human archive itself into art.

Bibliography

Beck, U. (1992). *Risk Society: Towards a New Modernity*. London: Polity.

Gibbons, P., Limoges, N., Nowotny, H., Schwarzmann, S., Scott, P. and Trow, M. (1994). *The New Production of Knowledge: The Dynamics of Science and Research in Contemporary Societies*. London: Sage.

Rees, M. (2003). *Our Final Century: The 50/50 Threat to Humanity's Survival*. London: Heinemann.

Tresilian, N. (2000). 'Atttractors and Vectors: The Nature of the Meme in Visual Art', in Ascott, R. (ed.). *Art, Technology, Consciousness –mind@large*. Bristol: Intellect Books, pp. 171–5.

Notes

1. Exstallation: I offer this coinage as a collective noun for all forms of large–format art produced outside a gallery or other art–dedicated site: artist placements, contextual works, earth art, land art, art–environments, happenings, performances, video networks and so on.

2. The Doppler effect in culture may explain why attempts to replicate anachronisms –as in 17th century Freemasonry, the witchcraft cults of the 18th century or the 19th century Gothic revival –tend to feed back as an uneasy pastiche in a limbo between 'old' and 'new'.

3 For the origins of the Mode 1 vs. Mode 2 science debate, see Gibbons, Limoges, Nowotny, Schwarzmann, Scott and Trow 1994.

5.10 Culture, Ecology and the Real

Paul O'Brien

The two cultures

In effect 'two cultures' have developed in uneasy coexistence: academic theory and political/ecological activism. On the one hand, academic instructors debate such issues as truth, identity and gender and wrangle about Derrida, Kristeva and Lacan. On the other hand, students organise politically around issues of globalisation, anti–capitalism and the environment, referring to Chomsky, Jeremy Rifkin and Naomi Klein.[1]

Today's alternative culture reaches back, in part, to older anti–capitalist ideas which post–modernism critiqued and apparently superseded –although the new politics is fonder of asking questions than offering pre–fabricated political remedies. In one manifestation, it dresses in colourful outfits and engages in a practice of 'tactical frivolity'.[2] Thus is reinstated the play principle with roots in Dada, Situationism and the Dutch 'Provos' of the 60s.

For its critique of industrialism, the new politics also has roots in the tradition of environmental philosophy, which is heavily influenced both by feminism and by the tradition of 'deep ecology' or ecosophy.

Feminism, ecology and science

Eco–feminism[3] is somewhat problematic within the context of post–modern thought, which radically doubts whether a term such as 'nature' has any fixed referent. The environmental standpoint is concerned with the limits of nature and the necessity to preserve it; the post–modernist or relativist view focuses on the cultural 'construction' of nature and its social role (Soper 1995 p. 7). Soper points out that it is not the discourse of 'global warming' or 'industrial pollution' that has created the conditions of which it speaks (ibid p. 249). She holds that myth, poetry and art are all grounds for believing that the delight and inspiration arising from nature indicate 'some relatively direct and unmediated responses to the environment' (ibid p. 244).

The critique of the humanity/nature, subject/object split harks back not just to the origins of western philosophy in Plato and Aristotle but in religious terms to a criticism of Genesis 1: 26–8, the notion of 'dominion'.[4] Consequently, the causes of our environmental problems may be seen as rooted in the religious and philosophical origins of western civilisation itself, in Judaea as well as Greece.[5] Historically, the metaphor of dominion spread from the religious to the social and political spheres (Merchant 1990, p. 3). This mindset was intensified by the scientific worldview that reconceived reality as a machine instead of an organism, thereby authorising the domination of nature as well as women (ibid p. xxi). The medieval theory of society had previously laid emphasis on the whole, while stressing the inherent value of each part:

> *The unity of the one was of higher value than the objectives of the many. The connection between the parts was integrated through a universal harmony pervading the whole. This organic cement bound together the macrocosm, the community, and the parts of each individual being or microcosm.*
>
> (ibid p. 71)

In an account of the 'scientific revolution', which exchanged the animistic notion of nature for a mechanistic one, Soper points out that:

> *This cosmological shift, in which a conception of nature as 'ensouled' organism is supplanted by a conception of it as inorganic, fundamentally mathematical and hence objectively quantifiable, has its correlate in the philosophical dualism of Descartes, which opposed God (the Architect and Prime Mover of the 'machine') to Nature (the 'clockwork' set in motion), and mind or soul (as the essence of humanity) to the body or inanimate matter of the rest of existence.*
>
> (ibid p. 43)

Merchant refers to Bacon's view that humanity lost its 'dominion' over creation as a result of the fall from the Garden of Eden, caused by a woman's temptation. The lost dominion could be recovered only by delving into the mine of knowledge about nature, through the interrogation of another female: nature conceived as such (Merchant 1990 p. 170). To illustrate the scientific methods, Bacon used the metaphor of the means used in the witch trials to extract information (ibid p. 168). Merchant notes Bacon's advocacy in his writings that nature be 'forced out of her natural state and squeezed and moulded'. In this way, in his terms, 'human knowledge and human nature meet as one'. This anticipates the experimental methods of today's scientists in terms of constraint of nature, anatomisation and the forcible discovery of hidden secrets –she notes Bacon's sexual imagery which foreshadows the scientific language of today, which lauds 'hard facts', a 'penetrating mind', the 'thrust of his argument' and so on (ibid p. 171).

In Bacon's utopian work New Atlantis, it was scientists who made the decision whether to reveal particular secrets or to keep them as the private property of their institute. The priest–like scientist of Bacon's depiction had, in his terms, 'an aspect as if he pitied men' (ibid p. 181). The parallel with today's scientists, guardians of information that they may or may not choose to share with the public, is evident (ibid p. 182).

Bacon described experiments in the New Atlantis:

> *By art likewise we make them greater or taller than their kind is, and contrariwise dwarf them, and stay their growth; we make them more fruitful and bearing than their kind is, and contrariwise barren and not generative. Also we make them differ in colour, shape, activity, many ways . . . We have also means to make divers plants rise by mixtures of earths without seeds, and likewise to make divers new plants differing from the vulgar, and to make one tree or plant turn into another.*

(quoted in ibid p. 183)

Merchant points out that a considerable part of Bacon's plan in the New Atlantis was aimed at dismantling prohibitions on manipulative magic (ibid p. 184). In fact, we are looking at the origins of the mindset that ultimately resulted in the culture of genetic engineering.

Merchant argues that the new view of nature as a female, the recipient of control and experimental dissection, served to sanction the exploitation of nature's resources. The Renaissance image of the nurturing earth gave way to new imagery of control (ibid p. 189). The new mechanistic order and the axiology of power associated with it would, in her terms, proclaim 'the death of nature' (ibid p. 190). A new notion of the rational self in a machine–like body came to replace the idea of the self as an integral part of society and the cosmos (ibid p. 214).

Drawing on the ideas of Heidegger in The Question Concerning Technology, she notes the power–orientation of post–Cartesian philosophy. In Heidegger's terms, modern technology has its essence in 'enframing', rendering nature as a 'standing reserve', a storehouse (ibid p. 228). Soper points out the strong elaboration of the Romantic critique in the argument of the critical theorists of the Frankfurt School: in its oppression of nature, 'instrumental rationality' severs us from nature as an origin (Soper 1995 p. 30).

The instrumental mindset of modern science is, then, traceable to a culture of domination endemic in western civilisation. The search for scientific 'truth' may not be value–free –perhaps it is already vitiated by the initial decision to subject nature to a dominating gaze. The basic question for Merchant and others writing from an eco–feminist perspective is whether the phenomenon of domination is ultimately cultural (i.e. specific to Judaeo–Christianity, western thought after Plato, and/or patriarchy) or fundamental to the human species itself.

Deep ecology

As Naess writes, deep ecology or 'ecosophy' involves rejecting the 'man–in–environment' model for the relational or total–field understanding. It also embraces the principle of biospherical egalitarianism (Naess 1990 p. 28). According to his view, if the 'self' is expanded in terms of breadth and depth, so that we conceive of the guardianship of free Nature as that of ourselves, then care is a natural result. 'Just as we need not morals to make us breathe . . . if your 'self' in the wide sense embraces another being, you need no moral exhortation to show care' (Naess 1995 p. 217). In Naess's terms, to act in a way that is ecologically fitting is, following Kant, to act beautifully –it is not a question of moral or immoral action (ibid). As Naess puts it:

> *So the norm 'Self–realisation!' is a condensed expression of the unity of certain social, psychological, and ontological hypotheses: the most comprehensive and deep maturity of the human personality guarantees beautiful action . . .We need not repress ourselves; we need to develop our Self.*

(Naess 1990 p. 86)

As Merchant points out, ecological ethics of this type transcends or relinquishes the distinction between 'is' and 'ought'. It is consequently mandatory for humans to live ecologically –they are part of the fabric of the environment that itself prescribes human behaviour and its bounds, since misuse will ultimately result in extinction. The environment itself determines our ethics in regard to it (Merchant 1990 p. 96).

Naess's ideas about deep ecology or 'ecosophy' predicated on human self–realisation have been criticised in terms of the difficulty of reconciling this with his bio–centric egalitarianism. There is a contradiction between biocentrism (which would include, perhaps, identifying with the mosquito or the AIDS virus) and human self–realisation. The alternative is a hierarchy of natural values in terms of richness, complexity, sentience, beauty and so on (Soper 1995 p. 257).

The basic question for deep ecologists like Naess is, what happens when there is a conflict between the interests of humanity on the one hand and nature on the other? It might, after all –according to the Gaia hypothesis –be in the interests of nature simply to get rid of us from the planet, given the damage we have already done. Is that something to which we should passively accede? The interests of humans may not always coincide with those of nature but it is arguable that they often or usually do.

Ecology and culture

The problems with developing an environmental ethic might be boiled down to the two questions: 1) Are we capable of living in an ecologically appropriate way? 2) Is it in our interests to do so? Given that the answer to both questions is a qualified 'yes', it would seem desirable that an environmental consciousness should come to permeate our culture.

The reality, though, is that the relationship between environmental values and contemporary culture is problematic. The art to the fore in the UK tends to bypass ecological issues in a downward–spiral search for shock value and ever–greater depths of philistinism. In Germany, on the other hand, a major ecological line stretches from Joseph Beuys to Anselm Kiefer.

There are a number of cultural practitioners making interesting work in the interface between art and biology,6 while some contemporary artists take note of issues such as nuclear power and environmental degradation.7 In the sphere of cyber–culture, the possibility is sometimes cited of creating at least a virtual version of the nature we (may) have ruined –an issue referenced in films from Soylent Green through Blade Runner to The Matrix.8

Ironically, the elements of an ecological critique seem strongest in the area of popular culture, traditionally criticised as dominated by commercialism and the culture industry. Despite the influence of Beuys, ecological consciousness has, until recently, largely bypassed the area of 'fine art', insofar as that term still has a meaning. This is perhaps due to a number of factors: an all–pervading ironisation of culture, a split between ethics and aesthetics going back for centuries and the marginalisation of 'romantic' concepts of natural beauty in contemporary theory and practice. There is also the post–modern suspicion of 'essentialist' notions of nature: a fixation, at least until fairly recently, on 'human' identity politics (particularly the politics of sexuality) and the replacement of the real by the simulacrum as a result of the writings of Baudrillard. There is the traditional suspicion of didacticism and the introversion and subjectivity often encouraged in art education. Finally, in political terms there is both the historical association of 'naturism' with fascism and the association of ecologism –particularly eco–feminism –with counter–cultural orthodoxies.

Perhaps, in the shadow of global catastrophe, environmental ethics will engage with aesthetics and ecological questions –if not necessarily answers –may come to infiltrate the culture of the intelligentsia, just as they have already infiltrated some areas of popular culture.9

Bibliography

Merchant, C. (1990). *The Death of Nature: Women, Ecology and the Scientific Revolution*. San Francisco: Harper and Row.

Naess, Arne (1990). *Ecology, Community and Lifestyle*. Rothenberg, David (trans. and ed.). Cambridge: Cambridge University Press.

Naess, Arne (1995). 'Self–realization: An Ecological Approach to Being in the World', lecture quoted in Fox, W. (1995). Toward a *Transpersonal Ecology: Developing New Foundations for Environmentalism*. New York: State University of New York.

Soper, K. (1995). *What is Nature?* Oxford: Blackwell.

White, L. Jr. (1967). 'The Historical Roots of Our Ecologic Crisis'. *Science 155*, pp. 1203–7.

Notes

1. Though there are elements –like eco–feminism –that contribute to both cultures.
2. This term might be used to describe the ludic activities of the 'Pink Block', also such recent counter–cultural phenomena as Etoy, the Yes Men, the Institute for Applied Autonomy and RtMark.
3. Some names in this respect are Carolyn Merchant, Val Plumwood, Mary Mellor, Karen J. Warren, Theresa Brennan and Verena Andermatt Conley.
4. The Hebrew 'radah' is even more violent, meaning treading down.
5. For a critique of the ecological crisis as being rooted in the ideology of western Christendom, see White (1967) pp. 1203–7. On the connection with Greece see the writings of H. Skolimowski and Morris Berman.
6. For example Eduardo Kac, Oron Catts, Ionat Zurr, Guy Ben–Ary, Joe Davis, Marta de Menezes, Karen Thornton, Eric Paulus, Malou Elshout, Dan Oki, Natalie Jeremijenko, Paul Perry, Mark Dion and art groups 'Gene Genies' and 'Clones–R–Us.'
7. For example Paul Fusco, Peter Fend, Cornelia Hesse–Honneger, Ann T. Rosenthal, Brandon Balengee and George Gessert.
8. One might mention also the implicit critiques of genetic engineering in Cronenberg's films The Fly and eXistenZ and, of course. there is Lisa (good) and Mr Burns (bad) in The Simpsons!
9. I am grateful to emerging artists Catherine Fitzgerald, Kate Minnock and Charlotte Swann for some references to artists above.

5.11 The Immersive Experience of Osmose and Ephémère: An Audience Study

Hal Thwaites

Introduction

It is still unknown what impact virtual reality (VR) experiences may have on the imagination and on the development of human consciousness, as discussed by researchers such as Heim (1993), Cartwright (1994), Attree et al. (1996) and Hansen (2001). To date, VR technologies have seen relatively widespread applications in education, research and medicine, therapy and rehabilitation, business and design and –especially –entertainment. On the other hand, very little investment has been made in VR applications that are designed to promote imaginative skills such as aesthetic sensitivity, reflection, artistic creativity or emotional insight. This lack parallels an equally important lack of research into fundamental questions about the *effects* of VR experiences and their impact on their audience. The immersive VR works of Char Davies are uniquely different from other VR applications in that they are designed specifically to facilitate processes of the imagination and they have been shown in museums to over 20,000 individuals worldwide, as described by Davis (1996), Pesce (2001) and most recently by Grau (2003). This research project on Davies's two works *Osmose* and *Ephémère* examines issues of the imaginative process, 'shifting awareness', consciousness as subjective experience and as an element of discovery while immersed, as described by Davies (1997, 1998). It also looks into the overall 'information impact' of the immersive VR experience (Thwaites 1991). A large majority of participants undergoing a visit to the works describe the experience as enriching, thrilling, inspiring and even rapturous, based on accounts from the Osmose Book of Visitor Comments, Museum of Contemporary Art, Montreal in 1995 as summarized by Treadwell (2002). It is from these departure points that the audience study of Osmose and Ephémère was conceived and designed. This project represents a major collaboration between an artist and a communication researcher in developing research tools to evaluate and better understand the audience/visitor experience within the gallery–art exhibit environment.

Owing to space limitations this paper can only present a preliminary analysis of the research data, starting with Osmose. The results of the greater project will be used by Davies in her research towards her new work and will contribute to the now small body of knowledge on the effects of artistic immersive virtual environments on the user–audience/participant. It is the first formalized research study on these two works.

The Study
The audience sample of the Osmose and Ephémère artworks comprised individual visitors to the John Curtin Gallery BEAP (Biennale Electronic Arts Perth) exhibition in Perth, Australia, where the works were on exhibition (Jones 2002). The study was conducted in the first two weeks of September 2002. Approximately 20 to 30 people per day could view Osmose, which alternated with Ephémère about half of the time. Over a period of 14 days, this made for a potential audience sample of between 140 to 210 subjects. The final audience sample consisted of 98 completed questionnaires, 44 from Osmose and 54 from Ephémère, taken from the visitors who volunteered to participate. The research questionnaire comprised 27 questions exploring the audience's reactions to, and feelings from, the immersion process. It was designed after consultation with Char Davies and John Harrison and a reading of the various museum comment books and the relevant literature. Questions included were both content and technology based, specific to the works and general on the field of VR overall. They were based on both cognitive and affective domains of the visitor experience, as described by Bruner (1991). Some of the areas covered were: emotional and physical feelings, sense of time and navigation, body awareness, overall enjoyment and recall of sounds and visuals. Subjects were also asked to draw their journey on a map of each work.

The project resulted in an incredible amount of data. The Osmose section alone generated approximately 1,188 discrete written responses and 41 immersion journey 'paths'. The responses were both Likert scaled items and open–ended replies. The latter exerted the least amount of control over the respondents and captured a wide variety of idiosyncratic differences. The questions were analysed both quantitatively and qualitatively. Individual differences by sex and age groups were also looked for. Specific trends and remarkable factors were looked for in each area and the similarities/differences of the works are being analysed.

Method of recruitment of participants
Visitors were asked if they would like to fill in the questionnaire after they had experienced the artworks on exhibit at the John Curtin Gallery. All visitors to the BEAP exhibit had to make appointments in advance to view Osmose, since it is an individual experience and requires a reserved time block of 30 minutes. The gallery organized and ran the performance of the work completely independent of this study. It was a voluntary informed activity, a fact that was clearly stated on the first page of the questionnaire.

Treatment of participants in the course of the research
This research activity was non–invasive, non–intrusive and of free consent. Subjects were informed of the nature of the research and that the study was being undertaken by Professor Thwaites, from Concordia University in Montreal. They were told that it would take about 10 to 15 minutes to complete the written

questionnaire, consisting of short answer or multiple choice type questions, that they were free to refuse if they so desired and that it was not required of them to participate. Since this exhibit was in Australia and open to the public, cultural differences were respected completely. Subjects were not paid as this study was on a 100 per cent completely voluntary basis. The questionnaires did not require names and there were only four questions that asked any personal demographic information: sex, age range, country of origin and occupation.

The research was thoroughly explained to each participant, both verbally and in writing. Participants were told that the anonymous results would be published. Individual or personal information was not asked for and basic demographic information was kept to a minimum. Each subject signed a release form. The research did not involve any follow-up procedure and was a one-time survey, in place, at the Curtin Gallery, in Perth.

Results

The Osmose visitor sample was composed of 44 subjects, 24 female and 20 male. They were mainly between the ages of 15 and 24 years (n=21) and 25 to 34 years (n=14). The remaining nine subjects were 35+ years of age. This was understandable, owing to the location of the BEAP exhibit within a university context. Overall there were no initial differences found between male/female responses to the majority of the questions. Further analysis is forthcoming.

The selected results of the ten Likert scale questions (multiple choice answers) and the seven Yes/No answer questions are presented in Table 1. Some questions combined both types of answers. The questions are numbered (from the questionnaire) and abbreviated in order to save space. Results are presented in a summary form. Missing subject responses account for the totals not always adding up to 100 per cent.

The open-ended type questions provided a rich amount of information on the effect on the audience of Osmose. Respondents were very articulate in describing what they experienced as summarised below in Table 2. Participants in the study were also asked to trace what they could remember of their immersion on a map/diagram of the Osmose world.

In addition to the written questionnaires, visitors were asked to volunteer for a video interview after their immersion. This facet of the project proved to be more difficult, since many people were less willing to be videotaped than to answer an anonymous questionnaire. However, approximately two hours of interviews were carried out over the two weeks of the project. Some of this material is available on the project website and will be reported on at a later date.

3. Did Osmose affect on you in any way?	91% Yes [% = 40/44]
4. How long did you feel you were immersed?	61% 15 min. or less, 20% no idea
5. Awareness of your body *while* immersed?	39% more, 23% same, 30% less
6. Awareness of your body *after* immersion?	43% more, 45% same, 7% less
12. Did you ever feel lost or confused?	48% once, 25% more than, 25% never
18. Did you remain aware you were in an art exhibit?	57% always, 36% start/end, 5% forgot all
19. Moving around inside Osmose was?	45% easy/very easy, 36% easy/hard to hard
20. Did you ever feel sick or uncomfortable?	72% no, 23% yes
21. Describe the head–mounted display.	57% uncomfortable, 43% adequate comfort
22. Is Osmose what you expected of a VR?	42% yes, 36% no
23. Have you ever experienced a VR before this?	73% no, 25% yes
24. Go back into Osmose, when would you go?	70% immediately–1 hour, 20% in a week

Table 1: Summary of Scaled and Yes/No responses (N=44)

The answers from the open–ended questions revealed a wide variety of idiosyncratic replies. Each questionnaire was entered exactly as written into a spreadsheet for archival and analysis purposes. As with most studies of this kind, it created a large body of text for coding. The responses were read and coded for common phrases, themes and variables. The instances of each were tabulated and converted to a percentage of the overall subjects, n=44. The data in Table 2 shows a summary of common grouped responses to the individual questions.

Conclusion

This paper represents the first report on a project using Char Davies's artworks as

1. How did you feel after Osmose?	25% Relaxed, 30% dreamy, 32% elated, 25% neg/sick
2. If you close your eyes and remember being inside Osmose, what do you see, feel or hear?	57% Sounds/insects, voices, birds, crickets, chords 55% Tree/leaves/forest / clearing, 32% Floating / tranquil calm/ airy, weightless, 27% Lights /dragonflies / fireflies
3. Has Osmose affected you?	20% Calm/refreshed, 18% dreamy, 16% sick/dizzy
7. Does using breath and balance for navigation contribute to the experience? 98% answered the question)	73% Positive: easy / heightened awareness / new experience/ effective/ control great/fantastic/ immersive, 23% Negative: hard to control/ frustrating /dizzy
8. Does the use of transparency contribute to your immersive experience, positively or negatively?	86% Positive: easy/ refreshing /weird/ softer feel/favourite part/loved it/newspace/freedom/ amazing/lack of boundaries, 16% Negative: less real/ disorienting / confusing / annoying
11. Places that caused emotion or physical sensation? (90% answered the question, 75% positively)	39% Lights, fireflies/wonderment, peace, joy, delight, 32% Underground /scary, unfamiliar, spooky, off balance, 20% Tree, forest, pond/awe, curiosity, exhilaration, floating
13. Part you enjoyed the most, why? (95% answered the question)	48% tree/leaves/forest, 32% moving /floating/passing through things, 20% calmness/beauty/pleasure, 16% lights/ dragonflies, 14% roots/code /text, 11% pond/abyss/ surfaces
14. Part you enjoyed the least? (90% answered the question)	27% HMD/navigation/lack interaction /resolution, 20% code/ text, 11% underground /roots, 9% trees/leaves/ stuck in place
15. Is there any aspect of Osmose that you unexpected or disturbing?	55% nothing, 20% lost way/HMD/ balance /out of time/music/darkness closing, 11% nausea/dizzy
16. What sounds did you enjoy the most, the least?	36% harsh, text/code, screeching/underground 30% forest, trees 18% birds/chirping 14% water, drops, pond, abyss, 11% codeworld, text, music, change of locations

Table 2: Summary of the open–ended responses (N=44)

a virtual reality 'test bed' as discussed by Lauria (1997). Many of Davies's personal reflections on the audience impact of her works and the reliability of the many quotes from the museum comment books are initially validated by this first group of subjects. Further analysis of the data is now being completed and will be reported at a later date in more detail and in comparison with the Ephémère data.

It was evident during the project that immersive virtual art environments are still a very new experience for the majority of visitors. Osmose had a definite emotional and cognitive impact on the audience as evidenced by the richness of the response data presented here. Visitors found rapture and pleasure in the immersive experience, which many were anxious to explore further and again. They remembered very small details in the imagery and soundscape. The major visual elements were predominant in the replies (the tree, leaves, pond, texts and lights). Davies's navigation system was a positive innovation for the users, giving them an increased awareness of their bodies while immersed, causing little or no instances of cybersickness. Most subjects found the artwork easy to explore and appreciated the transparent ambiguity of the imagery. It was also apparent that the HMD in current use was often cited as a negative factor in the overall experience. The subjects were willing to participate in VR research given a concise and simple research tool such as the questionnaire but they were less willing to do video interviews. The 'motion paths' proved to be a valuable addition to the questions in support of where immersants think or remember they were in the environments as a personal visit record. Overall the research subjects were articulate, detailed in their replies and willing to participate in this study on the immersive experience of Osmose.

Thinking is more interesting than knowing, but less interesting than looking.

(Goethe)

Bibliography

Attree, E.A., Brooks, B.M., Rose, F.D., Andrews, T.K., Leadbetter, A.G. and Clifford, B.R. (1996). 'Memory Processes and Virtual Environments: I Can't Remember What was There but I Can remem-

Figures 1–4: Sample Motion Paths for Osmose (enlarged versions on website)

ber How I Got There. Implications for people with disabilities. Proceedings of the First European Conference on Disability, Virtual Reality and Associated Technologies', Maidenhead, UK.

Bruner, J. (1991). *Acts of Meaning*. Cambridge, MA: Harvard University Press.

Cartwright, Glenn. (1994). 'Virtual or Real? The Mind in Cyberspace'. *The Futurist*, vol. 28, no. 2, pp. 22–6.

Davies, Char (1997). 'Changing Space: VR as an Arena of Being. Consciousness Reframed: Art and Consciousness in the Post–biological Era. Proceedings of the First International CAiiA Research Conference. Ascott, Roy (ed.). Newport: University of Wales College, Centre for Advanced Inquiry into the Interactive Arts.

Davies, Char (1998). 'Osmose: Notes on Being in Immersive Virtual Space'. *Digital Creativity*, vol. 9, no. 2, pp. 65–74.

Davis, Erik. (1996). 'Osmose'. *Wired Magazine*, vol. 4, no. 8, pp. 138–40, 190–2 .

Grau, Oliver (2003). 'Charlotte Davies: Osmose'. *Virtual Art, From Illusion to Immersion*. Cambridge, MA: MIT Press, pp. 193–211.

Hansen, Mark (2001). 'Embodying Virtual Reality: Touch and Self-Movement in the Work of Char Davies'. *Critical Matrix: The Princeton Journal of Women, Gender and Culture*, vol. 12, nos. 1–2, 'Making Sense', pp. 112–47.

Heim, Michael (1993). *The Metaphysics of Virtual Reality*. Oxford: Oxford University Press.

Jones, Stephen (2002). 'Breathing/Diving/Dreaming/Dancing BEAP in Perth (You Can't Buy These Emotions off the Hollywood Shelf)'. *Artlink*, vol. 22, no. 4 (New Museums/New Art), pp. 36–7.

Lauria, Rita (1997). 'Virtual Reality: An Empirical-Metaphysical Testbed'. *Journal of Computer Mediated Communication*, vol. 3, no. 2, pp. 1–25.

Pesce, Mark (2001). *The Playful World: How Technology is Transforming Our Imagination*. New York: Ballentine Books.

Thwaites, Hal and Malik, M. (1991). 'Toward Virtual Realities: A Biocybernetic View of Communication Media of the 21st Century', in Lasker, G., Koizumi, T. and Pohl, J. (eds.). *Advances in Information Systems Research*. Windsor, ON: The International Institute for Advanced Studies in Systems Research and Cybernetics, pp. 473–80.

Treadwell, Andrew (2002). *Virtual Transcendence* (Online) Oxford: Oxford Brookes University.

Notes

1. This project was made possible through a research grant from the Hexagram Institute in Montreal, Canada www.hexagram.org. Further details will soon be available on the Osmose Immersion Project Website www3.sympatico.ca/scooter5/ImerAud/study.html .

   It will also be linked to the Immersence website.

2. My sincere thanks to Char Davies for granting me access to her works for this project.

3. For a comprehensive listing of writings on both Char Davies' works, Osmose and Ephémère, please consult the Immersence Inc. website www.immersence.com/

4. I would also like to thank Professor Ted Snell, Curtin University Dean of Art;Patti Straker, Chris Malcolm, Jeff Khan, Cherie Duncan and the entire staff of the John Curtin Gallery, Perth, Australia, for their help in making my research stay so rewarding.

5.12 On Making Music with Artificial Life Models

Eduardo Reck Miranda

Introduction

Perhaps one the greatest achievements of Artificial Intelligence (AI) to date lies in the construction of machines that can compose music of incredibly good quality. One must not forget, however, that these AI systems are good at mimicking well-known musical styles. They are either hardwired to compose in a certain style or able to learn how to imitate a style by looking at patterns in a bulk of training examples. Such systems are therefore good for imitating composers or musical styles that are well established, such as medieval, baroque or jazz (Miranda 2000). Conversely, issues as to whether computers can create new kinds of music are much harder to study because in such cases the computer should neither be embedded with particular models at the outset nor learn from carefully selected examples. Furthermore, it is hard to judge what the computer creates in such circumstances because the results normally sound very strange to us. We are often unable to judge these computer-generated pieces because they tend to lack those cultural references that we normally hold onto when appreciating music.

One plausible approach to these problems is to program the computer with abstract models that embody our understanding of the dynamics of some compositional processes. Since the invention of the computer, many composers have tried out mathematical models, which were thought to embody musical composition processes, such as combinatorial systems, stochastic models and fractals (Dodge and Jerse 1985; Worral 1996; Xenakis 1971). Some of these trials produced interesting music and much has been learned about using mathematical formalisms and computer models in composition. The potential of Artificial Life models is therefore a natural progression for computer-generated music. By Artificial Life models here I mean those computational paradigms whereby the algorithms display some form of emergent behaviour that resembles a biological phenomena of some kind; for example, cellular automata models, generic algorithms, autonomous agents systems and adaptive games to cite but a few. This will become clearer in the next paragraphs where I introduce CAMUS and Chaosynth, two cellular automata-based music systems of my own design.

Listening to the propagation of Artificial Life patterns

CAMUS uses two simultaneous cellular automata to generate musical forms: the Game of Life and Demon Cyclic Space. Owing to limitations of space, we will briefly introduce only the role of the Game of Life in its generative process. More information on CAMUS can be obtained in a recent paper that appeared in the *Computer Music Journal* (McAlpine, Miranda and Hoggar 1999).

The Game of Life is a two–dimensional CA that attempts to model a colony of simple virtual organisms. In theory, the automaton is defined on an infinite square lattice. For practical purposes, however, it is normally defined as consisting of a finite $m \times n$ array of cells, each of which can be in one of two possible states: alive represented by the number one, or dead represented by the number zero. On the computer screen, living cells are coloured as black and dead cells are coloured as white. The state of the cells as time progresses is determined by the state of their eight nearest neighbouring cells. There are essentially four rules that determine the fate of the cells at the next tick of the clock:

- *Birth*: a cell that is dead at time t becomes alive at time $t+1$ if exactly three of its neighbours are alive at time t

- *Death by overcrowding*: a cell that is alive at time t will die at time $t+1$ if four or more of its neighbours are alive at time t

- *Death by exposure*: a cell that is alive at time t will die at time $t+1$ if it has one or none live neighbours at time t

- *Survival*: a cell that is alive at time t will remain alive at time $t+1$ only if it has either two or three live neighbours at time t

Whilst the environment, represented as E, is defined as the number of living neighbours that surround a particular live cell, a fertility coefficient, represented as F, is defined as the number of living neighbours that surround a particular dead cell. Note that both the environment and fertility vary from cell to cell and indeed from time to time as the automaton evolves. In this case, the life of a currently living cell is preserved whenever $2 \leq E \leq 3$ and a currently dead cell will be reborn whenever $3 \leq F \leq 3$. Clearly, a number of alternative rules can be set. The general form for such rules is ($Emin, Emax, Fmin$ and $Fmax$) where $Emin \leq E \leq Emax$ and $Fmin \leq F \leq Fmax$. The CAMUS implementation of the Game of Life algorithm enables the user to design rules beyond Conway's original rule. However, rules other than (2, 3, 3, 3) may exist but not all of them produce interesting emergent behaviour.

CAMUS uses a Cartesian model in order to represent a triple of notes. In this context, a triple is an ordered set of three notes that may or may not sound simultaneously. These three notes are defined in terms of the distances between them or intervals, in music jargon. The horizontal coordinate of the model represents the first interval of the triple and the vertical coordinate represents its second interval.

To begin the musical generation process, the CA is set up with an initial random configuration and is set to run. When the Game of Life automaton arrives at a live cell, its coordinates are taken to estimate the triple from a given lowest reference note. For example, if a cell at the position (5, 5) is alive then it will generate a triple of notes. The coordinates (5, 5) describe the intervals of the triple: a fundamental pitch is given, then the next note will be at five semitones above the fundamental and the last note ten semitones above the fundamental. Although the cell updates occur at each time step in parallel, CAMUS plays the live cells column by column, from top to bottom. Each of these musical cells has its own timing but the notes within a cell can be of different lengths and can be triggered at different times. Once the triple of notes for each cell has been determined, the states of the neighbouring cells in the Game of Life are used to calculate a timing template, according to a set of temporal codes. More information about this temporal codification can be found in a paper that appeared in *Interface* (now called *Journal of New Music Research*) (Miranda 1993).

Listening to organised chaos

Chaosynth is essentially a granular synthesiser (Miranda 1995; Miranda 1998a). Granular synthesis works by generating a rapid succession of very short sound bursts called granules (e.g. 35 milliseconds long) that together form larger sound events. The results tend to exhibit a great sense of movement and sound flow. This synthesis technique can be metaphorically compared with the functioning of a motion picture in which an impression of continuous movement is produced by displaying a sequence of slightly different images (sound granules in our case) at a rate above the scanning capability of the eye. So far, most of these systems have used stochasticity to control the production of the granules; for example, to control the waveform and the duration of the individual granules. Chaosynth uses a different method: it uses cellular automata. The CA used in Chaosynth tends to evolve from an initial random distribution of cells in the grid towards an oscillatory cycle of patterns.

The behaviour of this CA resembles the way in which most of the natural sounds produced by some acoustic instruments evolve: they tend to converge from a wide distribution of their partials (for example, noise) to oscillatory patterns (for example, a sustained tone). Chaosynth's CA can be thought of as a grid of identical electronic circuits called cells. At a given moment, cells can be in any one of the following conditions: *quiescent*, *depolarised* or *burned*. A cell interacts with its neighbours (four or eight) through the flow of electric current between them. There are minimum ($Vmin$) and maximum ($Vmax$) threshold values that characterise the condition of a cell. If its internal voltage (Vi) is under $Vmin$, then the cell is quiescent (or polarised). If it is between *Vmin* (inclusive) and *Vmax* values, then the cell is being depolarised. Each cell has a potential divider that is aimed at maintaining *Vi* below *Vmin*. But when it fails (that is, if *Vi* reaches *Vmin*) the cell becomes depolarised. There is also an electric capacitor that regulates the rate of depolarisation. The tendency, however, is to become increasingly depolarised with time. When *Vi* reaches *Vmax*, the cell fires and becomes burned. A burned cell at time t is automatically replaced by a new quiescent cell at time $t + 1$.

Each sound granule produced by *Chaosynth* is composed of several spectral components. Each component is a waveform produced by a digital oscillator (i.e. a lookup sampling table containing one cycle of a waveform) that needs two parameters to function: frequency and amplitude. The CA controls the frequency and duration values of each granule (the amplitude values are set up via another procedure). The values (i.e. the colours) of the cells are associated to frequencies and oscillators are associated to a number of cells. The frequencies of the components of a granule at time t are established by the arithmetic mean of the frequencies associated with the values of the cells associated with the respective oscillators. Suppose, for example, that each oscillator is associated with nine cells and that at a certain time t, three cells correspond to 110 Hz, two to 220 Hz and the other four correspond to 880 Hz. In this case, the mean frequency value for this oscillator at time t will be 476.66 Hz. The user can also specify the dimension of the grid, the amount of oscillators, the allocation of cells to oscillators, the allocation of frequencies to CA values and various other CA-related parameters. The duration of a whole sound event is determined by the number of CA iterations and the duration of the particles; for example, 100 iterations of 35 millisecond particles results in a sound event of 3.5 seconds' duration.

Discussion and conclusion

A number of professional pieces were composed using CAMUS-generated material, such as *Entre o Absurdo e o Mistério* (Miranda 2001) for chamber orchestra and the second movement of the string quartet *Wee Batucada Scotica* (Miranda 1998b). Chaosynth also proved to be a successful experiment in the sense that it is able to synthesise a large number of unusual sounds that are normally not found in the real acoustic world but nonetheless sound pleasing to the ear. As an example of an electroacoustic piece composed using the sounds of Chaosynth I cite *Olivine Trees*, which has recently been awarded the bronze medal at the International Luigi Russolo Electroacoustic Music Competition in Italy (the recordings of all these pieces are available on request). The results of the CA experiments are very encouraging, as they are good evidence that both musical sounds and abstract musical forms might indeed share similar organisational principles with cellular automata.

In general, I found that Chaosynth produced more interesting results than

CAMUS. I think that this might be due to the very nature of the phenomena in question. The inner structures of sounds seem more susceptible to CA modelling than large musical structures. As music is primarily a cultural phenomenon, in the case of CAMUS I think that one would need to add generative models that take into account the dynamics of social formation and cultural evolution. In this case, one should find modelling paradigms where phenomena (in our case musical processes and forms) can emerge autonomously. I am currently working on a number of experiments in which I am trying to simulate the emergence of musical forms in a virtual community of simple entities or 'software agents', in computer science jargon. Should such experiments corroborate my hypothesis that one can improve computer composition systems considerably by including mechanisms that take into account the dynamics of cultural evolution and social interaction, then I believe that over the next few years we will be listening to a new generation of much improved intelligent composing systems.

Bibliography

Dodge, T. and Jerse, T. A. (1985). *Computer Music: Synthesis, Composition and Performance*. New York: Schirmer Books.

McAlpine, K., Miranda, E. R. and Hoggar, S. (1999). 'Making Music with Algorithms: A Case–Study System'. *Computer Music Journal*, vol. 23, no. 2.

Miranda, E. R. (1993). 'Cellular Automata Music: An Interdisciplinary Project'. *Interface/Journal of New Music Research*, vol. 22, no. 1.

Miranda, E. R. (1995). 'Granular Synthesis of Sounds by Means of a Cellular Automaton'. *Leonardo*, vol. 28, no. 4.

Miranda, E. R. (1998a). *Computer Sound Synthesis for the Electronic Musician*. Massachsetts: Focal Press.

Miranda, E. R. (1998b). *Wee Batucada Scotica* (musical score). Amsterdam: Goldberg Edições Musicais.

Miranda, E. R. (ed.) (2000). *Readings in Music and Artificial Intelligence*. New York: Gordon and Breach/Harwood Academic Publishers.

Miranda, E. R., *Entre o Absurdo e o Mistério*, musical score. Amsterdam: Goldberg Edições Musicais, 2001.

Worral, D. (1996). 'Studies in Metamusical Methods for Sound Image and Composition'. *Organised Sound*, vol. 1, no. 3.

Xenakis, I. (1971). *Formalized Music*. Bloomington: Indiana University Press.

5.13 Artistic Strategies for Using the Arts as an Agent through the Creation of Hyper–Reality Situations

Karin Søndergaard

Using the Arts' as agents means to operate with a strategy in which one makes use of various formalized approaches to transform a situation into an artistic entity.

When you perform using formal strategies, alienation occurs.

An alien agent is a conceptual entity which, by methodically using a formal strategy, places itself in a state of trans–normality while simultaneously operating within normality. This alienation is established by having enhanced or 'hyper knowledge' of that specific structure within which one is going to act in such a way that with one's political statements, aesthetic corporal movements or social interrelations, it is possible to intervene in a more prepared manner and in much more detail, so that one not only has control of the situation by knowing what to do but likewise employs such a detailed state of preparedness that simultaneously, one is able to create and/or alter the situation in which one is acting.

Through this action it is possible to enter into a situation and influence its self–conception. You disturb the situation's self–perception and thereby provoke an attitude from all persons involved –politically, aesthetically or socially.

I will present three concrete projects that use such strategies and by doing so make the connection from Augusto Boal's 'Invisible Theatre' as political strategy, through formal choreographic movements as aesthetic strategy, to the use of a social strategy in an advanced interactive computer controlled installation with virtual robots.

The concept of 'The Invisible Theatre' as an artistic strategy of agency

In the 'Invisible Theatre', coined by Augusto Boal, the core approach is political intervention. When the political 'frames of discussion' starts to interpret and discuss themselves, a hyper–political situation occurs. Through this performative operation a progression is started –a trans–political situation –which makes one aware of the ambivalence and polemics in the construction of one's political position.

The function of the performers could be called hyper-personalities. They intervene in reality using actions designed specifically from the theme in question. To be able to raise and direct discussions on a specific subject, they prepare their opinions on the subject in detail and act purposefully. They reveal a certain complex of problems and provoke a confrontation with the 'audience' in a way that demands a position on the problem.

'Invisible Theatre' is a method of raising a political consciousness. It calls attention to the facts of repression and structures of power, as these are exposed in daily life.

The 'Invisible Theatre' is played at specific locations selected in relation to the theme and in relation to the audience one wishes to confront.

The 'Invisible Theatre' is invisible because the spectator is unaware of the fact that he is witnessing a performance. It is performed in a realistic style in the space of reality and thus experienced as part of ordinary reality.

The 'Invisible Theatre' confronts its audience with a number of scenes provoking them to participate. In this way, the spectators are transformed into co–actors –without actually realising that they are participating in a staged situation. The audience become active participants in the reality they experience, without knowing the fictional origin of the situation.

It is important to stick to the definition of 'Invisible Theatre' as theatre and that the performers play parts described in a manuscript. Furthermore, it is important that the actors are prepared on the variety of ways in which the situation might develop having the audience participate.

In order to be able to respond appropriately to the changing circumstances during the development of the play, the performers rehearse a collection of potential situations. In this way the performers prepare several possible trails of action and developments.

The strategic staging is the frame for the possible discussions and exchange of opinions. Within this frame the artistic intension unfolds. This unfolding could be called an agent –an agent that executes an agency –causing reflections and discussions on the given conditions.

One could say that the 'Invisible Theatre' constructs a reality within reality. It uses the artistic effects of theatre to cause a hyper–reality.

An agent provokes ambivalence in the given circumstances –the normality –generating new positions and new viewpoints as an artificial reality within reality.

The 'Invisible Theatre' transforms normality and something different –an otherness –emerges and asks questions and demand answers.

Using the political strategy of the 'Invisible Theatre' –in a café in 1979

This event was performed in a café often crowded with left-wing students. The theme was the student's self-recognised problems in communicating their political opinions. We wanted to stress the fact that being left wing and wishing to promote one's opinions was not only an issue of propaganda but also a question of personal openness.

> *the café was a crowded place with room for around 100 people. A group of four actors was placed at a table in the middle of the café. The actors represented members of a left–wing organization, having a meeting about how to reach 'the people' with their opinions. Two actors played newly arrived students seeking contact with the left–wing milieus. The two newcomers try to establish contact with the group but are rejected several times and the scene develops gradually into being loud and noisy, humiliating and embarrassing. In addition, two more actors discuss the incidence with the surrounding café guests, thus engaging them in the scene. The guests get involved in the discussions and the actors seek to guide the discussion toward the pre–planned theme*

Using an aesthetic strategy through formal choreographic movements

In the performance 'Glimt' (1989) we work with a formal intervention, through which a behavioural hyper–reality is established. It is a kind of slip –a trans–rituality which raises the awareness of the movements as rituals.

The focus of the work is formation of choreographic patterns in the public space. In the hectic, dense crowd of shoppers moving along the main pedestrian shopping street, we executed glimpses of choreographic movements. The formation of the choreographic patterns accentuates the ritual aspects of the movements of the crowd, by bringing attention to the pure formal qualities of movement. The performers act as some kind of behavioural agents, who we could define as *hyper–mannequins*.

The performers act like mannequins, who in repeated patterns of actions put themselves into positions in these figures. In this way the performers are the carriers of action–images. They execute movements in a progression of abstract logic that examines the movement as signs in indirect communication.

A sign could be a jump or a wave or it could be a longer composition of several signs, moved through space by the performer. The performers can respond to each other's signs, e.g. a jump can be responded to by a jump or the performer can join another performer's longer composition of signs. Patterns of signs emerge when several performers respond to each other's signs.

The signs do not have any meaning by themselves. They function as abstract codes and signals keeping the improvisation running and generating patterns of movements and signs.

Because the performers in their costumes are not differentiated from ordinary citizens and because the choreographically defined movements are executed in glimpses, the theatrical situation only becomes apparent momentarily.

The choreographic elements are rehearsed in the performers' laboratory as structures of improvisation –and can be thought of as an open and flexible entity. When this organizational structure of movements is transferred into the public space (which in itself functions as a partially organized flow of humans in motion), then it generates its own reality in the midst of the ordinary reality.

We are dealing with a presentation formally operating with a kind of otherness and which in glimpses throws normality into relief by going beyond the norm of behaviour. One could say that in this case the art operates as an alien agent forcing normality to obtain a new view of itself.

Reality is in this way used as a stage setting, available for reinterpretation.

Using a social strategy promoting interrelations between humans and virtual robots

In the interactive installation 'I Think You –You Think Me' the virtual entities Robert and Roberta establish a social system. We are dealing with a social space created by non–humans. One could say that it is a virtual social space controlled by two virtual representatives.

The installation is activated by human intervention into a virtual social space.

Robert and Roberta's conversation and mimicry generate the experience of a social system.

'Robert and Roberta' is a series of short animations and voices stored on a hard disc. The animations are files tucked into categories connected to the different states of minds or modes of Robert and Roberta. The voices are constructed from a similar set of categories, each containing a fixed number of pre–recorded sentences reflecting the state of mind. The voices are altered in real time with respect to parameters such as tempo and pitch in order to create an infinitive number of variations to the basic material.

A part of the software controls and decides the evolution of modes, through a continuous evaluation of the states of Robert and Roberta combined with the interaction sensor data from the visitors.

The installation is in this way choosing its expressions through a combination of the interaction of the audience and Robert and Roberta's own software controlled tendencies and their categories of sentences.

'I Think You –You Think Me' permits visitors to create relationships between real and virtual personalities. Two virtual beings, Robert and Roberta, are having a conversation. They are virtual entities in the sense that they are present with only their faces appearing on two computer screens while their voices emanate from loudspeakers. One might say that the 'real' life of these virtual beings then exists as bit–streams in the inner organs of the computers. Sensors permit Robert and Roberta to become aware of the movement and presence of real human beings in the room. When this happens, they speak directly to the visitors. However, when no one is in the room, they fall asleep and snore loudly. Through this set–up, visitors can join in and thereby extend Robert and Roberta's relationship with themselves through a series of questions and demands that are addressed to them. In doing so they enter into the ongoing discussion between Robert and Roberta and become involved in their world.

One might ask, 'Have we thought them up or have they thought us up?'

The focus of the installation is to create a three–way communication in the room.

One communication takes place in the inner minds of Robert and Roberta, their thoughts spinning around their own axis as they are meditating.

The second mode of communication enters as one of the two gets bored with thinking and starts a conversation with the other. The conversation will continue until either Robert or Roberta decides to rest and think for her/himself again. Robert and Roberta do not have minds like humans. They only try to mimic the behaviour, expressions and emotions of humans, which often render their conversations slightly absurd but at the same time having a recognizable character.

Finally, Robert and Roberta are sensitive to people who get close to them. Within a certain distance they will react to and turn their attention towards any visitor in their range. Inside their sensitive range Robert and Roberta will stop any ongoing conversation or meditation to start an interactive talk with the visitor (or alien, as Robert and Roberta call them).

In this installation reality is defined by virtual beings. The notion of reality is based on the virtual beings' behaviour –Robert and Roberta's sense of normality –and the audience will always be visitors in their reality.

When the audience enters this virtual social space they are becoming intruders in an already existing reality. They experience themselves as alienated and thus become increasingly aware of their own human nature and the human constructs of social relationships.

This is a kind of hyper–reality situation in which the virtual reality defines the cooperation with the real reality represented by the audience.

In this specific situation, the alien agent occurs as a disturbance of man's sovereignty.

Using the arts as an agent through the creation of hyper–reality situations

I think of an agent as the force that interprets the relationship between hyper–reality components. Hyper–reality is a matrix of understandings of the environment out of which emerge the possibility of an agent. Through artistic agency the organisation of the complexity of the reality crosses a threshold after which things no longer occur separately and unrelated. The artistic agency is forced into evolutionary patterns and forms a semiotic network of relations. This artistic agency can be understood as an agent.

From artefact to artistic impact

When thinking of art as an agent, the focus is moved from 'the work of art' to 'the impact of the artwork'. This means that the realization of art is moved from the level and manifests to the level of action and relationships.

An agent operates through its dynamic impact and causes the emergence of processes.

Bibliography

Boal, Augusto (1992). *Games for Actors and Non–actors*. London: Routledge.

Emmeche, Claus (1998). 'The Agents of Biomass', in Jürgensen, Andreas and Ohrt, Karsten (eds.). *The Mass Ornament*. :Denmark: Brandts Klædefabrik, pp. 65–79.

6. The Future
6.1 The Nanomeme Syndrome: Blurring of Fact and Fiction in the Construction of a New Science

Jim Gimzewski and Victoria Vesna

In both the philosophical and visual senses, 'seeing is believing' does not apply to nanotechnology, for there is nothing even remotely visible to create proof of existence. On the atomic and molecular scale, data is recorded by sensing and probing in a very abstract manner, which requires complex and approximate interpretations. More than in any other science, visualisation and creation of a narrative becomes necessary to describe what is sensed, not seen. Nevertheless, many of the images generated in science and popular culture are not related to data at all but come from visualisations and animations frequently inspired or created directly from science fiction. Likewise, much of this imagery is based on industrial models and is very mechanistic in nature, even though nanotechnology research is at a scale where cogs, gears, cables, levers and assembly lines as functional components appear to be highly unlikely. However, images of mechanistic nanobots proliferate in venture capital circles, popular culture and even in the scientific arena and tend to dominate discourse around the possibilities of nanotechnology. The authors put forward that this new science is ultimately about a shift in our perception of reality from a purely visual culture to one based on sensing and connectivity.

> *Micromegas, a far better observer than his dwarf, could clearly see that the atoms were talking to one another; he drew the attention of his companion, who, ashamed at being mistaken in the matter of procreation, was now very reluctant to credit such a species with the power to communicate.*

(Voltaire 1994 p. 24)

Introduction
Nanotechnology is more a new science than technology and the industry being constructed around it predictably uses old ideas and imagery. During its current rise to prominence, a strange propagandist 'nanometre' has emerged in our midst without being clearly realised by any of the participants. It is layered with often highly unlikely ideas of nanotech products that range from molecular sensors in underwear and smart washing machines that know how dirty the clothes are, to artificial red blood cells and nanobots that repair our bodies, all the way up to evil

swarms of planet–devouring molecular machines. Sensation–based media happily propagates this powerful and misleading cocktail combining scientific data, graphically intense visualisations together with science fiction artwork. In the past few years mixed–up nanomemes have emerged, where the differences between science fiction novels, front cover stories and images of reputable journals such as *Science* or *Nature* are becoming differentiated by the proportion of fiction to fact rather than straight factual content.

Venture capitalists, the military, governments around the world as well as educational institutions seduced by this syndrome are portraying nanotech as the saviour of our rapidly declining economies and outdated military systems. Dovetailing on the recent frenzied exponential rise and fall of information technologies, and also to a degree by biotechnology, the need for a new cure–all has been identified.

Two terms often used interdependently are nanoscience and nanotechnology. Surprisingly, the term nanotechnology predates nanoscience. This is because the dreams of a new technology were proposed before the actual scientific research specifically aimed at producing the technology existed. The term nanotechnology, in its short lifetime, has attracted a variety of interpretations and there is little agreement, even among those who are engaged in it, as to what it actually is. Typically, it is described as a science that is concerned with control of matter at the scale of atoms and molecules. 'Nano' is Greek for dwarf and a nanometre (nm) is one billionth of a metre, written in scientific notation as 1×10^{-9} m. Historically, the word nanotechnology was first proposed in the early 1970s by a Japanese engineer, Norio Taniguchi, implying a new technology that went beyond controlling materials and engineering on the micrometre scale that dominated the 20th century.[1]

One thing is certain, however –as soon as we confront the scale that nanotechnology works within, our minds short circuit. The scale becomes too abstract in relation to human experience. Consequently, any intellectual connection to the nanoscale becomes extremely difficult. Scientists have tried to explain this disparity by comparing the nanometre to the thickness of a human hair: the average thickness of a human hair is $\sim 5 \times 10^{-5}$m, which is (50,000) nm. Or, the little fingernail: around 1 cm across, which is equal to ten million nanometres. Recently, Nobel Laureate Sir Harry Kroto described the nanometre by comparing the size of a human head to that of the planet Earth –a nanometre would be the size of a human head in relation to the size of the planet if the planet were the size of the human head (Kroto 2002). But even that is difficult to intuitively grasp or visualise. What type of perceptual shift in our minds has to take place to comprehend the work that nanoscience is attempting and what would be the repercussions of such a shift? And how does working on this level influence the way scientists think who engage in this work? In our opinion, media artists, nanoscientists and humanists need to join forces together and envision such possibilities.[2]

On another level, as a metric, the nanometre itself does not do justice in describing nanotechnology but is rather the starting point of understanding complexity. Even the concept of precise fabrication at the ultimate limits of matter does nanotechnology injustice because it implies an industrial engineering model. When working on this kind of scale, we immediately reach the limits of rational human experience and the imaginary takes over. Researchers, science fiction writers and Luddites alike have gone into overdrive with the fantasies associated with the world driven by nanotechnology. One prevalent fear is mind control, while the dream is, as always, of immortality and power.

By some mysterious juxtaposition of events, the beginning of the 21st century is symbolised by the decoding of the genome, fears of distributed terrorist cells and nanotechnology as the big promise of total control of matter from the atom all the way up living systems. In the last ten years alone, over 455 companies based on nanotechnology have been formed in Europe, the United States and Japan, 271 major universities are involved in nanotech research and 95 investment companies are focusing on this new science. Over four billion dollars was invested globally in nanoscience in 2001 and the bar is being raised

(Taylor 2002; Bainbridge and Roco 2002; Tolles 1994; Holister and Harper 2002). But, unlike infotech and to a degree, biotech, nanotech is very much in its infancy of development and principally in the research phase. Perhaps this is what makes it so attractive to such a varied audience –the field is wide open for visionaries and opportunists alike, representing new uncharted territory resembling the early stages of space exploration of the 20th century and mission–oriented approaches to science and technology. Indeed, NASA foresees this potentially disruptive technology as being instrumental in exploring space to answer such questions, as 'Are we alone in this universe?'[3]

Although nanotechnology is used widely to refer to something very tiny, this new science will eventually revolutionise and impact upon every single aspect of our lives. It will do this on all scales all the way up from the atom to the planet Earth and beyond. The very *modus operandi* of science is already changing under its influence. Nanoscience not only requires input from practically every scientific discipline but it also needs direct and intense collaboration with the humanities and the arts. It is highly probable that this new technology will turn the world, as we know it, upside down, from the bottom up.

Richard Feynman is often credited as the person who initiated the conceptual underpinnings of nanotechnology before the term was coined. Although many physicists who were working in the quantum realm arrived at perhaps similar conclusions, his lecture, 'There is Plenty of Room at the Bottom', given in 1959, is used as a historical marker for the conceptualisation of nanoscience and technology. Indeed, it is interesting to note that this was not an invention *per se* but more a shift of focus or attention generated by a flamboyant personality that is interpreted to initiate the advent of nanotechnology.[4]

Many of Feynman's visions really took hold in the early 1980s when nanoscience and technology truly took off. In 1981 Heinrich Rohrer and Gerd Binning, at the IBM Zurich research laboratories, invented the Scanning Tunnelling Microscope (STM). which for the first time 'looked' at the topography of atoms that cannot be seen (Binning 1999). With this invention the age of the immaterial was truly inaugurated. Not much later, in 1984, a molecule was discovered by Sir Harry Kroto, Richard Smalley and Robert Curl that really got the ball rolling. 'Buckminsterfullerene' was named after Buckminster Fuller, an architect, engineer and philosopher whose dome structures employed geometries found in natural structures (Applewhite 1995). Not coincidentally, the IBM PC was taking centre stage and causing a true revolution in arts and sciences alike. In a short period of history, many new things appeared, creating a perfect environment for a natural symbiosis between science, technology and art. Another decade would pass before people occupying these creative worlds would expand their perceptual field to include each other's points of view. Indeed, the surge of this expansion happened from a genuine need to embrace and cross–pollinate research and development between science, technology and art.

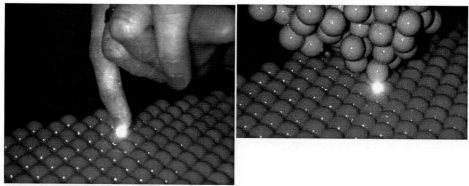

Figure 1: Principle of a scanning tunnelling microscope uses a local probe. The gentle touch of a nanofinger is shown in (a) where if the human finger was shrunk by about ten million times it would be able to feel atoms represented here by spheres 1 cm in diameter. If the interaction between tip and sample decays sufficiently rapidly on the atomic scale, only the two atoms that are closest to each other are able to 'feel' each other as shown in (b) where the human finger is replaced by an atomically sharp tip. Binning and Rohrer (1999) inspired this explanation of the STM.

New vision: the STM –a symbol of the shift from visual to tactile perception

Up until the mid–1980s, scientists viewed matter, atoms, molecules and solids using various types of microscopes or in abstract space (Fourier Space). The widespread use of optical microscopes had begun in the 17th century, enabling people like Galileo to investigate matter through magnification by factors of hundreds. These microscopes relied on lenses and the properties of light as a wave. Waves were manipulated by lenses to magnify and create an image in the viewer's eye, providing information on how light is reflected or transmitted through an object.[5]

Typically, human perception of a microscope is a tube–like structure through which one looks and sees reality magnified. In a deeper philosophical sense, while being strictly scientific, the concept of 'seeing' is illusory. Nevertheless, when one looks through a microscope at a butterfly's wings, it is difficult to separate one's conscious mind and its interpretation from the information transmitted by one's eyes.

Figure 2: The STM records images of surfaces and molecules as a two–dimensional data set of heights. Here an ordered array of molecules called hexa–butyl decacyclene, each around 1 nanometre in size, were recorded by the STM. The resulting data was then plotted as a grey scale image representing the apparent height of the molecules. Each molecule is represented as six lobes in a distinct hexagonal pattern with a dark central portion. Interestingly this height map does not represent the real height of the atoms but rather the probability of parts of the molecule to convey electrons by quantum tunnelling to the tip. The casual observer tends to see the pattern as representing the shape of the molecule (Gimzewski et al.: unpublished data).

The eye itself contains a small part of the brain that already pre–processes the information received as light particles or waves. As the magnifying power of the microscope increases, the average person looking through the lenses maintains his or her illusion of seeing a reality and interprets the image in terms of common human experience related to the scale in which one normally observes the world.

The Scanning Tunnelling Microscope[6] represents a paradigm shift from seeing, in the sense of viewing, to tactile sensing –recording shape by feeling, much like a blind person reading Braille. The operation of a STM is based on a quantum electron tunnelling current, felt by a sharp tip in proximity to a surface at a distance of approximately one nanometre. The tip is mounted on a three–dimensional actuator like a finger, as shown schematically in figure 1. This sensing is recorded as the tip is mechanically rastered across the surface producing contours of constant sensing (in the case of STM this requires maintaining a constant tunnelling current). The resulting information acquired is then displayed as an image of the surface topography (fig. 2). Through images constructed from feeling atoms with an STM, an unconscious connection to the atomic world quickly becomes automatic to researchers who spend long periods of time in front of their STMs. This inescapable reaction is much like driving a car –hand, foot, eye and machine coordination becomes automated. Similarly, the tactile sensing instrument soon became a tool to manipulate the atomic world by purposefully moving around atoms and molecules and recording the effect, which itself enabled exploration of interesting new physical and chemical processes on an molecule–by–molecule basis[7] (fig. 3).

Figure 3: View of a Scanning Tunnelling Microscope (STM) at the PICO lab of one of the authors (Gimzewski) at UCLA

In science, commonly agreed human perceptions are constantly in question. Indeed, as the power of the 20th–century microscopes increased, the images recorded progressively reflected not only patterns of waves determined by physical object form but also how the light waves scatter and interfere with each other. The butterfly's blue wings no longer have colour –one finds the colour to be an illusion –a beautiful illusion –where form, shape and periodic patterns on the nanoscale manipulate light waves to provide us with the illusion of seeing blue (Ghiradella 1991). As the magnification increases, we can no longer rely on our common human perception. Rather we see how, in this case, nature has carefully duped us –how through some magnificent evolutionary process, she has generated what is called nanophotonics (Yablonovitch 1993). Nanophotonics is a way to manipulate light through shapes, not mirrors. Indeed, by just changing the physical structure of matter on the nanoscale, we can produce a mirror, a mirror that is perfect; a mirror that some time in the future, through voice command, will switch to become a window. As we increase magnification into the truly invisible realm, we change our perception to view the world around us as an abstraction, a pattern of light waves. We apply mathematical principles based on fundamental rules for the way light intensifies with itself and object form. From this analysis comes an interpretation, perhaps as a mathematical reconstruction of reality.

Both nanotechnology and media arts, by their very nature, have a common ground in addressing the issues of manipulation, particularly sensory perception, questioning our reaction and changing the way we think. They are complementary and the issues that are raised start to spill over into fundamental problems of the limits of psychology, anthropology, biology and so on. It is as if the doors of perception have suddenly opened and the microscope's imperfection of truly representing object form forces us to question our traditional (western) values of reality.

Magnification –on the edge of reality
Scientists progressively turned up the magnification but no matter how good the glass lenses were, how precise the brass tubes and screws were, at around x10,000, the image goes fuzzier until, at x100,000, the image is basically blank. This is called 'the Raleigh limit', which says you cannot see anything by using a wave that is smaller than half the size of the wave. In other words, this light wave has a size just like an ocean wave has the distance between its crests. The length of the wave is the feature that limits what we 'see' which has a limit when we use regular light of 200 nanometres. It is already twice the size of a wire in a Pentium 4 or a few hundred times thinner than a hair –it is back to the metric. To get higher magnification, scientists used shorter waves and even the wave properties of electrons. Nevertheless, despite the progress, the high energies and conditions required to make these higher resolution images started to destroy the very objects they wanted to 'see'. In effect, they ended up looking at matter using something like a focused blowtorch in a vacuum.

During the early 1980s a dramatic moment happened in microscopy that has led to the rapid growth of nanoscience. It was a simple idea that put the whole concept of lenses into disarray. An IBM team, Henrich Rohrer, Gerd Binning, Christoph Gerber and Eddie Weibel were working on finding pinhole defects in nanometre–thin oxide layers that acted as barriers for quantum tunnelling for what was known as the Josephson project. Pinholes as tiny as a nanometre shorted out the tunnelling process. These were difficult to characterise using traditional microscopes and the researchers used a tiny needle to contact the oxide layer to probe the electrical properties of the film.

Necessity as the mother of invention prompted the researchers to build a machine where the little needle could be moved across the surface of the oxide film and thereby seek out the pinholes. Thinking about how the new invention worked led to an important realisation scribbled on a lab book. By using electron tunnelling as a probe, they realised that mainly the last atom on the needle (the closest on the

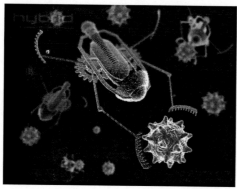

Figure 4: Floating in an aliquot of laboratory test fluid, these hypothetical early medical nanorobots are testing their ability to find and grasp passing virus particles. Courtesy of Jeff Johnson, 2001. Copyright 2003 Hybrid Medical Animation.

surface) really sensed the local properties. The needle moving across the surface created a topographic representation and a few back–of–the–envelope calculations indicated even single atoms could be resolved. What they realised was a major paradigm shift –rather than using lenses and waves, they were recording by feeling. In 2003 a whole range of microscopes based on tactile sensing have now been developed and many companies were established to manufacture machines that are used by microelectronics and data storage industries. Worldwide sales of these machines are in the range of a billion dollars.

The new tactile techniques opened up a radically new approach to microscopy enabling real local properties to be imaged and mapped. For instance, ultra high–resolution images of local magnetism like bits of north–and south–directed domains could be obtained with magnetic tips. If friction were an issue, images of local friction as it scanned the surface could be mapped. This opened up a new world, a world never really seen before on those terms –the nanoworld. Even bigger consequences of 'touching' rather than looking were also realised.

The environment in which one could image at really high resolving power with the *wave* microscopes was limited to a vacuum so that biological objects were dead and perturbed from their natural state. This was a major drawback that limited the interpretation of microscope images of biological samples. The new tactile microscopes were not subject to such limitations. Consequently, it was possible to image *in vivo* what is under physiological conditions on live specimens. It was possible to image the electrodes of batteries as they worked in their acidic environment. Our windows on the nanoworld looked not at parts of systems but really at operating fully functional systems, allowing their complexity to be 'seen' and measured all the way up from the nanoscale of individual atoms and molecules.

Through the paradigm shift in microscopy, the tactile probes which were now being called 'scanned probe microscopies' (SPM) opened up yet one more feature, which had been the 'Holy Grail' of a mad dream. The idea to manipulate and move single atoms and molecules in a controlled manner, up until the mid–1980s, was something outside of general scientific plausibility. In fact, the famous scientist

Erwin Schrödinger wrote (1952 p. 233) that we would never experiment with just one electron, atom or molecule. This view, like concepts of statistical mechanics that viewed matter as a collective property of atoms and molecules, required a rebel to question its almost religious doctrine. Eight years later, Richard P. Feynman told us that there are no physical limitations to arranging atoms the way we want but he was pretty much alone when he said it (1960 p. 22).

The mechanist view: molecular nanobots

He was living like an engineer in a mechanical world. No wonder he hadbecome dry as a stone.

(Simone de Beauvoir 1954, The Mandarins)

The invention of clocks is commonly accepted as the beginning of our move away from natural cycles. The Benedictine monks invented the clock to regulate the time of devotional prayers and it was not long before it was incorporated in the town hall to be used by shops and merchants. But the mechanical age does not take full hold on the western tradition until the mid–1600s, with the Industrial Revolution. Thomas Hobbes in 1651 reduced the body to mere mechanical parts, followed by Descartes who placed humans outside of nature and firmly established the dualism that is still very strong in our society and in the world of science in particular. To Descartes, the universe functioned as a giant machine, according to mathematical laws that can be unravelled and controlled. Everything external to the human is there for the human to use and manipulate. This philosophical stance was put to action in science by Newton who extended the machine analogy to all laws of nature. Newtonian physics describes everything from the motion of stars to atomic particles as part of the mechanical structure. Newtonians held that by discovering these laws, men prove their superiority and affirm their natural right to dominate all of matter in nature. During this same time, the English philosopher and statesman Roger Bacon argued (in support of science) that the purpose of the mechanical arts was to yield profit and societies were formed that funded scientific ventures focused on solving navigational and military problems. There is much more that could be said about the political and economic climate of Europe at this time and how this story has led us to the world we face today but for the purpose of this paper we will only allude to this connection. Suffice it to say that the clock speeds of computers and the magnification levels of microscopy today far exceed the capabilities of the human biological system and that clearly we have reached the limit of the mechanical age. This has led some to believe that machines are superior to humans and that robots will take over the world.

The term robot, as is well known, was first used in a science fiction book entitled *Valka s Mloky* (*War with the Newts*) and a play entitled *RUR* (*Rozuma Universalni Roboti –Rozum's Universal Robots*; *rozum* means wisdom). written by Karel Capek in 1920. Capek coined the word 'robot' from a Czech word *robota*, meaning drudgery or compulsory labour. This idea of the robot was assimilated into the science world and developed to the point where robots rove the terrain of Mars and vacuum homes. Perhaps the most controversial persona in the robotic sciences is Hans Moravec, whose visions no doubt compete with science fiction writers. Moravec, a professor of robotics at Carnegie Mellon University, envisions a not–too–distant future in which robots of superhuman intelligence, independent of their human creators, colonise the planets in outer space. He projects that in the next 40 years of robot development there will be a rise of super–intelligent, creative, emotionally complex cyber–beings and the end of human labour. He predicts an absolutely mechanistic future trajectory in which robot corporations will reside in outer space and he imagines planet–size robots that cruise the solar system looking for smaller robots to assimilate. Eventually every atom in the entire galaxy would be transformed into robotic space, with a full–scale simulation of human civilisation running as a subroutine as depicted in the 1999 movie *The Matrix*.

These mechanistic ideas are not in the realm of the physical only but have equally been distributed by the cyber realm in the version of a robot. 'Bot', an abbreviation for robot, came into use with the development of autonomous software programs that typically run in the background on the Internet. The first popular version of bot was used in MUDs and MOOs, online social spaces where fact and fiction commonly blur.[8] Many different types of bots emerged, depending on the focus of the program –adbot, knowbot, pokerbot, searchbot, smartbot, spambot, to name a few. It is this train of thought from which the idea of a 'nanobot' emerged –commonly visualised as derivatives of robots on a nanoscale.

In the introduction to *Engines of Creation* (1986). Drexler turns to a dictionary definition of a machine as 'any system, usually of rigid bodies, formed and connected to alter, transmit, and direct applied forces in predetermined manner to accomplish a specific objective, such as the performance of useful work' and asserts that molecular machines fit this definition quite well. Throughout the book, natural systems are interpreted as machines operating to Newtonian principles and the nanomeme is firmly established in an mechanical engineering worldview. Proteins, ribosomes, RNA, DNA and viruses are all part of a grand machine. In 1992 Drexler, who is an engineer by degree, went beyond predicting a general emergence of nanotechnology and wrote another book –*Nanosystems: Molecular Machinery, Manufacturing and Computation* –detailing technical particulars. Drexler's drawings of nanothings tend to resemble molecule–sized versions of mechanical counterpoints that have been around since the Industrial Revolution: gears, cogs, levers and pistons (Gao et al. 1996). If these versions of nanomachines will some day materialise, his engineer's calculations, which hold true in the world most people comprehend, will probably not be of much use in the molecular realm. However, his futurist vision includes self–assembled armies of tiny robots that build greater armies of tinier robots, *ad infinitum*.

The creation of this nanomeme that is currently in circulation has certainly been promulgated by Drexler's work and later by the Foresight Institute Inc., a futurist organisation based in Palo Alto. The concept of nanotechnology in the general public stems in part, from the media's habitual reliance on the promotions and prognostications of Drexler and his Foresight Institute. The mechanistic nanomeme has taken on the sheen of authority, as one press clipping breeds another. Indeed, the nanomeme is similar to the self–replication of the nanobots themselves. Many articles have an inspired tableaux of molecule–sized robots 'grabbing atoms one by one' and then replicating armies of themselves. The Foresight Institute website asserts a lot of things, such as –within the foreseeable future, there will be a 'nanobox' that manufactures items such as cell phones from a 'toner' composed of 'electrically conductive molecules' –and so on. In the long run, we will turn dirt into food, ending world hunger, which is another theme that propagated around some nanotechnology enthusiasts who believe it will give humans the power of telepathy (see www.foresight.org/).

During the 1950s and 1960s strategic thinking using 'systems analysis' emerged, pioneered by the Rand Corporation, a military research and development institution. This was happening at the same time that the greatest discovery in biology occurred –the physical structure of DNA. Watson and Crick explicitly described DNA in computer terms as the genetic 'code', comparing the egg cell to a computer tape. This school of thought was perpetuated in even more extreme terms by proponents of Artificial Life such as Chris Langton, who spoke of separating the 'informational content' of life from its 'material substrate' (Langton 1989). As Richard Coyne noted: 'Information is thought to be the essence of life, as in the DNA code. To record and break the code is to have mastery over life' (Coyne 1995 p. 80).

In 1995 the Rand Corporation published a study on the potential of nanotechnology (Anton 2001). The Rand paper relied heavily on the writings of Drexler and the Foresight Institute. The authors concluded that nanotechnology would best be used to take advantage of indigenous resources found on asteroids, comets or planets for mining; defending Earth against impacts; or tools to assist extensive colonisation of the solar system on a reasonable timescale. Interestingly, ending wars, hunger or solving the energy crisis gets no mention at all.

Weaving fact and fiction into blur

If you want to think of it that way, a human being is actually a giant swarm. Or more precisely, it's a swarm of swarms, because each organ –blood, liver, and kidneys –is

> a separate swarm. What we refer to as 'body' is really a combination of all these organ swarms. We think of our bodies as solid, but that's only because we can't see what is going on at the cellular level. If you could enlarge the human body, blow it up to a vast size, you would see that it is literally nothing but a swirling mass of cells and atoms, clustered together into smaller cells and atoms.
>
> (Crichton 2002a p. 260)

Michael Crichton's recent novel, *Prey*, provides an excellent example of contemporary science fiction that is based on current science. As an acclaimed author of best-selling novels that are almost always converted into blockbuster movies, there is no doubt that he influences the collective imaginary. *Prey* was almost immediately on the New York Times top five bestseller list and remained there for weeks. Rights for a movie were bought by Twentieth Century Fox before the book was completed and in September 2002 the author signed a partnership agreement with SEGA to develop a game scheduled to be released in 2004. Crichton takes four separate fields: distributed processing for networked computing; nanotechnology or molecular manufacturing; biotechnology and the behavioural science of socially organised insect communities, such as bees and ants. By tying in the evolution process, he comes up with a very plausible scenario and some possible 'Particle Swarm Organisation' applications. As he says in his foreword: 'Sometime in the twenty-first century, our self-deluded recklessness will collide with our growing technological power.'

Crichton includes in his novel many references to current histories and skilfully weaves them into the story, blurring fact and fiction. For instance, the main character who narrates the story describes how scientists (Don Eigler and co-workers) at IBM repositioned Xenon atoms with an STM tip to form the letters of the company logo (Eigler 1990). In his narrative he also comments that this was more of a stunt than anything else and that it would take much more to create new technology. The description of the building of molecular assemblers in the book is directly inspired by Drexler's visions of nanobots. After laying down a foundation based on actual events, the author proceeds to tell the story of a company that succeeds in building molecular assemblers that eventually go out of control.

After referencing events, people and companies many of us are familiar with, we are taken on a horror ride that instils a real fear of nanotechnology. This and many other works of science fiction that have appeared in the movies, TV programmes, books and PC games reflect the concept of nanotechnology as 'more' than science or hard technology. It has actually evolved into a culture and art form in its own right. Even more than cyborgs, AI or robots, nanotechnology truly traverses science and art as the dream of the future. This novel will first be read by millions, then will be watched in movies and played in games, until it finally becomes another part of the collective nanoconsciousness.

It is also notable that scientists have honoured Crichton by actually naming a new species of dinosaur after him. Scientists at the Institute of Vertebrate Palaeontology and Palaeoanthropology of the Chinese Academy of Sciences named a new ankylosaurus species 'Crichtonsaurus bohlini' in honour of Crichton and Birger Bohlin, a Swedish palaeontologist. The three–yard–long fossil of an armour–plated vegetarian, discovered in northeast Liaoning Province, is estimated to date from 90 million to 100 million years ago (Crichton 2002b).

Dark visions

In 1752 Voltaire wrote a diminutive science fiction story about microbes, men and beings from outer space. The story, entitled *Micromegas* (from the Greek small/great). was written while he was living at a humanist retreat dedicated to the science of Newton and the philosophy of Locke. The philosophers are visited by aliens who are introduced to a world that is hugely affected by the new discovery of microscopy. The microscope in Voltaire's book, similar to the STM today, challenged accepted reasoning and belief systems. Amazingly, in this story, we have also fiction based on the science of its day but set in a time plagued by wars. Specifically, during the writing of *Micromegas* the Russo/Austrian–Turkish war was underway (1736–9) and the aliens are told that these new discoveries about matter will be used for evil, for that is what men do:

> *We have more than enough matter to do plenty of evil, if evil comes from matter; and too much of evil, if evil comes from the spirit. For instance, do you realise that as I speak a hundred thousand lunatics of our species, wearing helmets, are busy killing and being killed by a hundred thousand other animals in turbans, and that everywhere on Earth this is how we have carried on since time immemorial?*

(Voltaire 1994 p. 30)

Public conceptions of nanotechnology and the blurring of fact and fiction seem to go hand in hand more than in any other science. As nanoscience is being established, it is clear that the imagination is there to roam the many dark visions connected to the military's interest in nanotech and soon we are also in the midst of a new type of 'war' that will not only require new tactics but also new technologies. Almost immediately after the 9/11 World Trade Center attack there were numerous scares of the use of anthrax, with many news reports speculating on the use of biological weapons.

Anticipating such a scenario, the US Army is also collaborating with MIT, having recently promised the university $50 million for a new Institute for Soldier

Nanotechnologies (ISN). The aim is to improve soldiers' protection and their ability to survive using new tiny technologies to detect threats and automatically treat some medical conditions. The army isn't the only branch of the military actively developing smart textiles. The US Navy funded a project in 1996 that eventually turned into the Smart Shirt, a product commercialised by SensaTex Inc. in Atlanta, with technology from the Georgia Tech Research Corporation. The T–shirt functions like a computer, with optical and conductive fibres integrated into the garment. It can monitor vital signs, such as the heart rate and breathing of wearers, and will most likely be first put to use by law enforcement officers and military personnel (Kary 2002).

From nanobots to nanobods

> *Wisdom requires a new orientation of science and technology towards the organic, the gentle, the non–violent, the elegant and beautiful.*
>
> *(Schumacher 1973 p. 27)*

Does it really make sense to extend the idea of a mechanical robot to software program bots and apply a Newtonian/ industrial–age approach to work on the molecular level? More and more researchers working on this scale are looking closely at natural biological systems for clues and inspiration. In this vision of bio–inspired nanotechnology, the body and mind shift to another paradigm and certainly appear much more appropriate to the new century we have just entered.

Schumacher, an economist, in *Small is Beautiful*, maintains that the prevalent pursuit of profit and progress, which promotes giant organisations and increased specialisation, has in fact resulted in gross economic inefficiency, environmental pollution and inhumane working conditions. With the emphasis on 'person not product', he points the way to a world in which capital serves people instead of people serving capital. Around the same time, Buckminster Fuller, an engineer, moved away from the mechanistic view by studying carefully natural systems (that also appeared on the molecular scale with the later discovery of the C60 molecule) (Fuller 1975). Recently, Smalley, one of the Nobel Laureates who discovered buckminsterfullerene, responded to the well–established nanomenes by pointing out that the main problem that the world currently needs to address is the energy crisis. He stopped short of connecting nanotech to this problem but he certainly has made a significant attempt to shift the discourse of the current hype.[9]

> *With the increasing computing power, research of the invisible realm increased at an accelerated pace at the end of the 20th century. In addition to the decoding of the human genome and the discovery of a new form of carbon, a new type of life is also being found. One such recently discovered creature that does not fit into any previous category of life was found in an undersea vent north of Iceland. These creatures, for-*

mally known as Nanoarchaeum equitans, may represent an entirely new grouping within archaea, the most mysterious of life's three domains. They are small spheres attached to other organisms and are so genetically strange and so tiny –smaller than a grain of sand and about the width of four human hairs –that they were invisible to traditional ecological survey methods. Even the ultimate molecular ecology methods could not detect these new microbes because they are so different from everything known so far. Karl Stetter, a professor of microbiology at the University of Regensburg in Germany who led the discovery team, and his colleagues detected the creatures only after growing them in hot, oxygen–free and high–pressure conditions to simulate their natural hostile environments. The DNA of the 'nano–sized hyperthermophilic archaeon' is interesting because it is so minimal –containing just 500 kilobases. The genome is among the smallest known to date. There are 6 million kilobases for humans and 9 million for corn. (The others are eukaryotes –organisms with nucleated cells like people, plants and fungi –and bacteria) (Huber 2002 p. 63).

We should take a closer look at ourselves as magnificent nanobeings, connected and part of an entire living body of this Earth and beyond for inspiration, not to machines of the past. DNA, proteins and cells of all sorts already function at nanoscale in animals and plants and they work at normal temperatures. In our view, the nanobots of the past with their mechanical structures, batteries, motors and so on are evolving into 'nanobods' –a closer reflection of our human condition in which living nanoscale chemical–mechanical elements are connected in ever–increasing complexity along the principle of cells, the smallest general unit of life capable of autonomous replication.

Conclusion: nano fact and fiction: being in–between

Nanotechnology works at a scale where biotech, chemistry, physics, electrical and mechanical engineering converge and thus has real potential to impact upon every aspect of our lives. We will see an impact on everything from our social systems to buildings, furniture, clothes, medicine, bodies and minds. But most of all, where we believe it will make a fundamental shift is in our conscious and unconscious minds. As the perception of reality shifts to the collective level, we will find ourselves in an entirely new world, with very different values and motivations. However, we do acknowledge that any radical proposition, with such enormous and global implications, will undoubtedly have to face fierce opposition from those who have so much invested in the old, mechanistic worldview. We have witnessed, in the 20th century, many great innovations that have been squashed by corporate, industrial and national interests –transportation and energy being at the top of the list. It appears to us that resistance to a technology that will change fundamentally the way humans think may be much greater, given the usual time period of 20 to 50 years it takes for technology to penetrate into general society. We are about to witness some great ideological struggles, much greater than what we have seen in past centuries. Indeed, the stage has already been set for this new era with the basic moral rights to own one's genetic code exemplified by the patenting of genes and the cloning of human beings.

In nanotechnology, the blurring of fact and fiction is very much part of the developing narrative in the construction of a new science and industry. This blurring is not necessarily negative and has a potential to connect media arts, literature and science in many new and interesting ways. Art, literature and science working together is certainly a powerful combination that should be nurtured in education on all levels. As common technologies are being used in arts, sciences and practically all disciplines, borders are becoming increasingly indiscernible and we have to be more conscious than ever of the metaphors being generated. The barriers between disciplines and people in them are more or less psychological. Currently, the vast majority of stories and imagery being circulated in the public realm are based on 20th–century thinking that is largely centred on machines. Nanoscale science and media art are powerful synergies that can promulgate the 21st–century emergence of a new third culture, embracing biologically inspired shifts, new aesthetics and definitions.

Bibliography

Anton, P.S. (2001). *The Global Technology Revolution: Bio/nano/materials Trends and Their Synergies with Information Technology by 2015*. Santa Monica, CA: Rand.

Applewhite, E.J. (1995). 'The Naming of the Buckminsterfullerene'. *The Chemical Intelligencer*, vol. 1, no. 3 (July).

Bainbridge, M.C. and Roco, W.S. (eds.). (2002). *Converging Technologies for Improving Human Performances: Nanotechnology, Biotechnology, Information Technology and Cognitive Science*. NSF/DOC sponsored report. Arlington, VA. June.

bama.ua.edu/~hsmithso/class/bsc_656/websites/history.html.

Binning, G. and Rohrer, H. (1999). 'In Touch with Atoms'. *Rev. Mod. Phys.* 71, pp. S324.

De Beauvoir, S. (1954). *The Mandarins*. New York: Norton and Company.

Besenbacher, F. (1996). 'Scanning Tunneling Microscopy Studies of Metal Surfaces'. *Rep. Prog. Phys.*, 59, pp. 1737–802ff.

Bradbury, S. (1968). *The Microscope Past and Present*. London: Pergamon Press.

Brown, P.D., McMullan, D., Mulvey, T. and Smith, K.C.A. (1996). 'On the Origins of the First Commercial Transmission and Scanning Electron Microscopes in the UK'. *Proc. Royal Microscopical Society*, vol. 21, no. 32 , p. 161.

Capek, K .(1970). *RUR and Insect Play*. New York: Pocket Books.

Capek, K. (1936 [in Czech]). *War With the Newts*. Trans. M. and R. Weatherall. Berkeley Medallion Edition paperback, May 1967; Catbird Press paperback, March 1990; Northwestern University Press paperback, October 1996.

Coyne, R. (1995). *Designing Information Technology in the Postmodern Age*. Boston: MIT Press.

Crichton, M. (2002a). *Prey*. New York: Harper Collins.

Crichton, M. (2002b). 'World Briefing | Asia: China: Science Nods To Dinosaur Fiction'. *New York Times*, 11 December, p. 8.

Crommie, M.F., Lutz, C.P. and Eigler, D.M. (1993). 'Imaging Standing Waves in a Two–dimensional

Electron Gas'. *Nature,* 363, p. 524.

Cuberes, M.T., Schlittler, R.R. and Gimzewski. J.K. (1996). 'Room–Temperature Repositioning of Individual C60 Molecules at Cu Steps: Operation of a Molecular Counting Device'. *Appl. Phys. Lett.* vol. 69 no. 20, pp. 3016–18.

Drexler, K.E. (1986). *Engines of Creation*. Garden City, NY: Anchor Press/Doubleday.

Drexler, K.E. (1992). *Nanosystems: Molecular Machinery, Manufacturing, and Computation*. New York: John Wiley & Sons.

Eigler, D.M., Lutz, C. P. and Rudge. W.E. (1991). 'An Atomic Switch Realised with the Scanning Tunneling Microscope'. *Nature*, 352, p. 600.

Eigler. D.M. and Schweizer, E.K. (1990). 'Positioning Single Atoms with a Scanning Tunneling Microscope'. *Nature*, 344, p. 524.

Feynman, R.P. (1960). 'There's Plenty of Room at the Bottom'. *Sci. Eng.*, 23, p. 22.

Feynman, R. (1993). 'Infinitesimal Machinery'. *J. Microelectromechanical Sys*, 2, pp. 4–14.

Fuller, R.B. (1975). *Synergetics Dictionary*. New York: Macmillan Press.

Gao, G., Cagin, T., Goddard III, W.A., Globus, A., Merkle, R. and Drexler, K.E. (1996). 'Molecular Dynamics Simulations of Molecular Planetary Gears', in *First Electronic Molecular Modelling & Graphics Society Conference*.

Ghiradella, H. (1991). 'Light and Color on the Wing: Architecture and Development of Iridescent Butterfly Mirrors'. *Applied Optics*, 30, pp. 3492–500.

Gimzewski, J. (1997). 'Atoms Get a Big Push, or is That a Pull?'. *Physic World*, November, pp. 27–8.

Gimzewski, J. and Joachim, C. (1999). 'Nanoscale Science of Single Molecules Using Local Probes', *Science*, 283 (5408). pp. 1683–8.

Holister, P. and Harper, T.E. (2002). *The Nanotechnology Opportunity Report*. CMP Cientificia, No. 1 and 2, March.

Huber, H. et al. (2002). 'A New Phylum of Archaea Represented by a Nanosized Hyperthermophilic Symbiont'. *Nature*. 417, pp. 63–7.

Hunter, E. (1993). *Practical Electron Microscopy: A Beginner's Illustrated Guide*. Cambridge: Cambridge University Press.

Jung, T.A., Schlittler, R.R., Gimzewski, J.K., Tang, H. and Joachim, C. (1996). 'Controlled Room–Temperature Positions of Individual Molecules: Molecular Flexure and Motion'. *Science*, 271 (5246). pp. 181–4.

Kary, T. (2002). 'MIT to Make "Nanotech" Army Wear'. CNET News.com, 14 March.

Kroto, H. (2002). 'Nanoeterscale Architecture', in *The Proceedings of The Second International Symposium on Nanoarchitectonics Using Suprainterationcs* (NASI2). 26–8 March 2002. Los Angeles: University of California Press.

Langton, C. (1989). *Artificial Life*. Redwood City: Addison–Wesley.

Moravec, H. (1986). 'Book Review: *Engines of Creation* by Eric K. Drexler'. *Technology Review*, vol. 89, no. 7 (October). pp. 76–7.

Schrödinger, E. (1952). 'What is Life'. *Br. J. Philos.*, 3, p. 233.

Schumacher, E.F. (1973). *Small is Beautiful, Economics As If People Mattered*. New York: Harper and Row.

Stroscio, J.A. and Kaiser, M.J. (eds.). (1993). *Scanning Tunneling Microspectroscopy*, 27. San Diego: Academic Press, Inc.

Taylor, J.M. (2002). *New Dimensions for Manufacturing: A UK Strategy for Nanotechnology*. London: Department of Trade and Industry.

Tolles, W.M. (1994). *Nanoscience and Nanotechnology in Europe*. Navel Research Laboratory,

NRL/FR1003–94–9755, 30, December.

Voltaire, F. (1994). *Micromegas and Other Short Fictions.* New York: Penguin Classics.

Whitfield, J. (2002). 'New Bug Found on Bug: Marine Microbe Sets Miniaturisation Records'. *Nature, Science Update,* 2 May.

Yablonovitch, E. 'Photonic Band–gap Crystals [Condensed Matter]'. *J. Phys.*, 5, p. 2443.

Notes

1. Norio Taniguchi of Tokyo Science University first defined nanotechnology in 1974 (Taniguchi 1974). 'On the Basic Concept of "NanoTechnology"', *Proc. Intl. Conf. Prod. Eng.*, Part II, Tokyo: Japanese Society of Precision Engineering.

2. To address this need, the authors, together with Katherine Hayles, have joined forces and created SINAPSE, a 'non–centre' that is devoted to promoting collaborations of creative thinkers in arts, sciences and humanities. In November 2002, together with UC DARNET (Digital Arts Research Network). SINAPSE co–sponsored a conference entitled 'From Networks to Nanosystems' that was attended by media artists connected to the CAiiA–STAR programme and scientists from UCLA. See http://sinapse.arts.ucla.edu and Martin 2002 pp. 3–5).

3. NASA Ames Research Center has advanced computational molecular nanotechnology capabilities to design and computationally test atomically precise electronic, mechanical and other components and work with experimentalists to advance physical capabilities; see: www.cmise.ucla.edu/.

4. Feynman was known to challenge authority and caused consternation in his years with the Manhattan Project, which developed the atomic bomb, by figuring out in his spare time how to pick the locks on filing cabinets that contained classified information. Without removing anything, he left taunting notes to let officials know that their security system had been breached. In 1965 Feynman was awarded the Nobel Prize, along with Shinichero Tomonaga of Japan and Julian Schwinger of Harvard University. The three had worked independently on problems in the theory of quantum electrodynamics, which describes how atoms produce radiation. He reconstructed almost the whole of quantum mechanics and electrodynamics, deriving a way to analyse atomic interactions through pictorial diagrams, a method that is still used widely.

5. For more information on the history of microscopy, see Bradbury (1968). Hunter (1993). Brown, McMullan, Mulvey and smith (1996). For a history of optical and electron microscopes, see: bama.ua.edu/~hsmithso/class/bsc_656/websites/history.html.

6. For more information on the STM, see Besenbacher (1996) and Stroscio and Kaiser (1993).

7. For more information on the molecular manipulation using the STM, see: Crommie, Lutz and Eigler (1993 p. 524); Eigler and Schweizer (1990 p. 524); Eigler, Lutz and Rudge (1991 p. 600); Gimzewski and Joachim (1999 pp. 1638–88); Gimzewski (1997 pp. 27–8); and Cuberes, Schlittler and Gimzewski (1996 pp. 3016–18).

8. A MUD or a 'Multi–User Dungeon' is a virtual social experience on the Internet, managed by a computer program and often involving a loosely organised context or theme, such as a rambling old castle with many rooms or a period in history. MUDs existed prior to the World Wide Web, accessible through telnet to a computer that hosted the MUD. A MOO is an Object Oriented MUD, i.e. the programming language allows you to create 'objects' and follows the same principle. Today, many MUDs and MOOs can be accessed through a website and are better known as '3–D worlds'.

9. Professor Richard E. Smalley made this statement at the 16th Annual Glenn T. Seaborg Symposium, 26 October 2002 in a lecture, 'Bandgap Fluorescence from Buckytubes'.

Contributor Biographies

Robert Pepperell
Robert Pepperell is an artist and writer. He studied at the Slade School of Art and went on to work with a number of influential multimedia collaborations including Hex, Coldcut and Hexstatic. As well as producing experimental computer art and computer games he has published several interactive CD–Roms and exhibited numerous digital installations including at the Glasgow Gallery of Modern Art; the ICA, London; the Barbican Gallery, London; and the Millennium Dome, London.

Steve Grand
Steve Grand was a NESTA Dreamtime Fellow during 2003–4 and holds honorary fellowships in Psychology at Cardiff and Biomimetics at Bath, but is otherwise an independent scientist, artist or engineer, according to interpretation. Website: ww.cyberlife–research.com

Andrea Gaugusch
Ph.D. in Philosophy, University of Vienna, 2001; M.A. in Psychology, University of Vienna, 1999. Lives in Vienna. Her research interests are Cognitive Science, Philosophy of Science, Philosophy of Virtual Reality, Music Psychology. Andrea Gaugusch was an Advanced Research Associate of the Planetary Collegium, University of Plymouth. 2003–2004.

Michael Punt
Michael Punt is Reader in Art and Technology at the University of Plymouth. He is also Editor in Chief of *Leonardo Digital Reviews*, a member of the Leonardo/ISATS Advisory Board and the MIT/Leonardo Book Series Committee.

Nina Czegledy
Nina Czegledy, an independent media artist, curator and writer, has been involved in collaborative international projects for over a decade. Czegledy's interest in art, science and technology is reflected by projects in progress including the Aurora Effector, BioSense and Electromagnetic Bodies.

Paul Newland, Chris Creed, Maestro Ron Geesin
Paul Newland, Chris Creed, Maestro Ron Geesin are Senior Research Fellows in the Responsive Environments Centre, School of Art, Design and Media, University of Portsmouth.

Armando Montilla
Armando Montilla is a Venezuelan architect and theorist with professional formation and work experience in Canada and the US. Since 2000 he has done post–graduate research work at the Architectural Association in London and at the Bauhaus Dessau Foundation in Dessau, Germany, where he participated in the *Bauhaus Kolleg* Research Programme.

Evgenija Demnievska
Evgenija Demnievska is Art Director of the Co–Resonance association, Montreuil. She has exhibited in over 40 one–woman shows and participated in more than 200 group exhibitions and in symposia. She works on interactive and multidisciplinary projects with the participation of artists and of the public (fax and Internet art projects).

Roy Ascott
Roy Ascott is Director of the Planetary Collegium, University of Plymouth and also Visiting Professor in Design/Media Arts, University of California, Los Angeles. His previous publications include *Telematic Embrace: Visionary Theories of Art Technology and Consciousness* (University of California Press 2003).

Oron Catts and Ionat Zurr
Oron Catts and Ionat Zurr are artist/researchers and curators. Oron Catts is the Co–founder and Artistic Director of SymbioticA – The Art & Science Collaborative Research Laboratory at the School

of Anatomy & Human Biology, University of Western Australia. He also founded the Tissue Culture and Art Project in 1996, and exhibits and publishes internationally. Ionat Zurr is artist–in–residence at SymbioticA, and co–founder of the Tissue Culture and Art Project. She is a PhD candidate researching the ethical and epistemological implications of wet biology art practices, and publishes internationally.

Yacov Sharir
Yacov Sharir is a doctoral researcher in the Planetary Collegium at the University of Plymouth. He is a frequent keynote speaker at arts and technology conferences and symposia in the USA and around the world. Sharir is a contributor to numerous international publications, journals and books related to issues regarding international interdisciplinary art and technology

Kjell Yngve Petersen
Kjell Yngve Petersen is a Lecturer in performance composition and new technology in the School of Arts, Brunel University, London. He is an artist, the artistic and administrative director of Boxiganga, Copenhagen., and doctoral candidate at CAiiA, in the Planetary Collegium, University of Plymouth.

Gordana Novakovic
Gordana Novakovic's artistic language is inspired by theoretical physics, biology and mathematics. Based on similarities among elements of micro and macro cosmos, her work consists of universal symbols, organic and mathematical forms. She is currently researching the potentials of expanding and merging different digital networks into an experience of mixed realities.

Anthony Crabbe
Anthony Crabbe is leader of the Design Contract Research Unit at Nottingham Trent University. From a background in critical studies, he has moved increasingly into product design and development, both as a project leader and as a designer.

Geoff Cox and Adrian Ward
Geoff Cox and Adrian Ward work together on a number of projects including *Vivaria.net* http://www.vivaria.ne and are both trustees of the UK Museum of Ordure http://www.museum–ordure.org.uk. They have previously written 'The Aesthetics of Generative Code' (with Alex McLean) and 'The Authorship of Generative Art' http://www.generative.net/.

Eril Baily
Eril Baily lectures at Sydney College of the Arts, University of Sydney, and has for the past four years been Associate Dean of Postgraduate Studies. She gained her doctorate in philosophy and has subsequently focussed her research on Consciousness in an Electronic Environment.

Margaret Dolinsky
Margaret Dolinsky is an Assistant Professor and Research Scientist at Indiana University in Bloomington. Dolinsky researches, designs and creates CAVE Automatic Virtual Environments. She is a PhD candidate at CAiiA in the Planetary Collegium, University of Plymouth.

Garth Paine
Garth Paine is a freelance composer, sound designer, installation artist and academic. He has been commissioned extensively in the United Kingdom, Germany, the USA and Australia, producing original compositions and sound designs for over 40 film, theatre, dance and installation works in the last ten years.

Ioanna Spanou and Dimitrios Charitos
Ioanna Spanou is currently pursuing a two–year Masters degree in Landscape Architecture at the Technical University of Catalonia (Universidad Politecnica de Catalunya). She is a practising architect and has received a number of awards in architectural competitions in Greece.

Dimitrios Charitos is a lecturer in the Department of Communication and Media Studies of the University of Athens. He was awarded a Diploma of Architecture (NTUAthens 1990), an MSc in CAAD (1993) and a PhD in 'The Design of Virtual Environments' from the Department of Architecture, University of Strathclyde, Glasgow.

Stahl Stenslie
Stahl Stenslie is working on the development of experimental interface technologies and tools within the fields of art, media and network research. He lives and works as a media artist, curator, scientist and media researcher in Oslo, Norway and Cologne, Germany. He is presently working with cognition and perception manipulative projects as well as undertaking PhD studies in New Media.

Maia Engeli
Maia Engeli is an information architect, specialising in the design of information access and exchange. She combines digital networks, computer graphics and artificial intelligence to create information and communication environments that supplement human talent and cognitive skills.

Mauro Cavalletti
Mauro Cavalletti is an Interaction Design Director at R/GA, a New York based design studio, working on the development of interactive communication. In recent years he has designed interaction experiences, information architecture and user interfaces for large-scale global web sites for clients such as IBM and Ericsson.

Adriana de Souza e Silva
Adriana de Souza e Silva is a PhD candidate at the School of Communications at the Federal University of Rio de Janeiro (UFRJ), Brazil. Since August 2001 she has been a visiting scholar at de Department of Design Media Arts at UCLA. Her research, 'Hybrid Space Nomads', studies the impact of nomadic technology devices, like cell phones, in the organization of urban spaces.

Mike Phillips and Chris Speed
The Arch–OS project is managed by the Institute of Digital Art and Technology and produced by members of STAR and CNAS research groups based in the University of Plymouth. Arch–OS is produced in collaboration with the architects and engineers: Feilden Clegg Bradley Architects, Buro Happold, Nightingale Associates, Hoare Lea, DrMM (Derijke Marsh Morgan), Signwave/CASM. Arch–OS is a collective of individuals working from the School of Computing at the University of Plymouth. Mike Phillips (Director of i–DAT) and Chris Speed (Tele–Social Navigation/Spaceman) are here representing the Arch–OS development team.

Christina McPhee
Christina McPhee, a transmedia artist, lives in California. Christina studied painting with Philip Guston at Boston University. In 2002 physical installations of www.naxsmash.net showed at Convergence, New Media Centre for Cybersonica, Institute of Contemporary Art, London; and at FILE Electronic Language Symposium, Sao Paulo; as performance installation.

Claudia Westermann
Claudia Westermann was born in 1971 in Heidelberg, Germany. She studied architecture at the University of Karlsruhe, Germany, and the University of Tampere, Finland, achieving a diploma in architecture. She has worked as an architect and is an official member of the German Chamber of Architects.

Shaun Murray
Shaun Murray is a practising architect in London, and a PhD candidate with the Planetary Collegium.

Dene Grigar
Dene Grigar is an Associate Professor of English at Texas Woman's University and specializes in new media, interactive arts, electronic literature, rhetoric, and Greek literature and culture. She is Associate Editor of *Leonardo Reviews* and International Editor for *Computers and Composition*. In 2001 she attended an NEH Summer Seminar at UCLA led by N. Katherine Hayles, an experience that led her to undertake, from 2002–4, a post–doctoral study with the Planetary Collegium.

David Topping
David Topping is an artist–researcher based in South Wales, UK. He is an Associate Senior Lecturer with the Communication, Media and Culture Group at Coventry School of Art & Design, Coventry

University. He is currently undertaking a PhD at Exeter School of Art and Design, University of Plymouth, titled Information Art www.informationart.net

Donna Cox
Donna Cox is Professor in the School of Art and Design, University of Illinois, Urbana–Champaign; and Division Director for Experimental Technologies, National Centre for Supercomputing Applications. Cox's computer graphics and visualisations are featured in art and science museums, television and IMAX theatres around the world.

Diana Reed Slattery and Charles Rene Mathis
As Associate Director of the Academy of Electronic Media at Rensselaer Polytechnic Institute, Diana Slattery researches, designs, and produces highly interactive, game–like multimedia environments for education, entertainment, and the arts. Her Ph.D. research in visual language and interactive narrative informs the Glide project.

Charles René Mathisis a Master's candidate in Computer Science at Rensselaer Polytechnic Institute. He is the programmer for LiveGlide. Charles is currently working on BiblioGlide, a unique set of research tools combining a database and a 3D visualization interface, using the LiveGlide forms.

Eduardo Kac
Eduardo Kac is a Chicago–based artist and writer and is Professor and Chair, Art and Technology Department.

Ernest Edmonds
Ernest Edmonds works in the constructivist tradition and first used computers in his practice in 1968. He is currently is Professor of Computation and Creative Media at the University of Technology, Sydney where he runs a multi–disciplinary practice–based art and technology research group, the Creativity and Cognition Studios. He has exhibited throughout the world, from Moscow to LA and published widely in the area.

Paul Brown
Paul Brown is an artist and writer who has been specialising in art and technology for 30 years. He is currently (2002–5) a Visiting Fellow at Birkbeck College, University of London, where he is working on their CACHe project (Computer Arts, Contexts, Histories, etc.). Examples of his artwork and publications are available on his website at: www.paul–brown.com

Clive Myer
Clive Myer is Director of the Film Academy at the University of Glamorgan and Executive Committee member of the National Association for Higher Education in the Moving Image. He was Founder Director of the International Film School Wales, Programme Leader in Film and Television at the University of London Goldsmiths College and Executive Committee member of the Independent Filmmakers' Association. He has been an independent filmmaker for 30 years.

Jane Tormey
Jane Tormey lectures in Critical & Historical Studies at Loughborough University School of Art & Design, UK. She co–edits the electronic journal *Tracey –Contemporary Drawing* and has work published in *Masquerade: Women's Contemporary Portrait Photography* (Ffotogallery 2003), *The State of the Real* (I.B. Tauris 2005), *IJADE* and *AfterImage*.

Daniel Meyer–Dinkgräfe
Daniel Meyer–Dinkgräfe is a lecturer at the Department of Theatre, Film and Television Studies, University of Wales, Aberystwyth. He has numerous publications on the topic of *Theatre and Consciousness* to his credit and is founding editor of the peer–reviewed web–journal *Consciousness, Literature and the Arts* www.aber.ac.uk/tfts/journal.html.

Niranjan Rajah
Niranjan Rajah assistant professor in School of Interactive Art and Technology, Simon Fraser University.He is a doctoral candidate at CAiiA, Planetary Collegium, University of Plymouth He was Associate Professor and Deputy Dean at the Faculty of Applied and Creative Arts, University Malaysia

Sarawak where he lectured from 1995 to 2002. He was visiting professor at the department of Design / Media Arts, UCLA in 2001.

Julia Rice
Julia Rice is currently acting head of the media department, and teaches media studies, at St Cyres School in Penarth. Her current research includes the convergence of technologies, both in the historical and futuristic perspectives. She is also interested in researching film at PhD level and is an active member of several academic list serves.

Alex Shalom Kohav
Alex Shalom Kohav is an installation artist, poet and writer on art, consciousness and Kabbalah. He is the author *That Annoying State Between Naps: A Kabbalistic History of Consciousness and Ten Riddles of the Hebrew Sphinx: Consciousness and the Kabbalah of Moses*. His artwork has been exhibited at the National Center for Atmospheric Research (NCAR), Boulder, Colorado; UCLA; the Museum of Neon Art (MONA), Los Angeles; BMoCA (Boulder Museum of Contemporary Art); Documenta X, Kassel, Germany (as part of Thing.Net exhibit); and Ft. Collins, Colorado MOCA.

Ron Wakkary
Ron Wakkary is an Associate Professor at Simon Fraser University in the programme of Information Technology and Interactive Arts and a doctoral candidate in the CaiiA–STAR programme at the University of Plymouth.

Nicholas Tresilian
Nicholas Tresilian is an art historian, broadcaster and founding director of UK media PLCs in the UK and Europe. He is currently Vice–President of the US–based International Society for the Study of Time. He lectures and writes on the relationship between art–history and cultural evolution.

He was for many years a director of Artist Placement Group and its successor Organisation and Imagination.

Paul O'Brien
Paul O'Brien teaches aesthetics, cyber–culture and art theory in the National College of Art and Design, Dublin. He is particularly interested in the interface between aesthetics, politics, culture and technology. His publications are in the areas of art, culture and new media.

Hal Thwaites
Professor Thwaites holds degrees in Communication Arts and Educational Technology with a ten–year professional background in television production at the CBC. Thwaites's research and teaching is in media production, multimedia, information design, virtual reality and 3D media. rofessor Thwaites is also President of the International Society on Virtual Systems and Multimedia (VSMM) and Chair of the VSMM 2003 International Conference.

Eduardo Reck Miranda
Eduardo Reck Miranda is a Reader in Artificial Intelligence and Music at the University of Plymouth, UK. He is a leading research scientist and a composer of international reputation. He has an MSc in Music Technology (University of York) and PhD in Music (University of Edinburgh) His research has made important contributions in the fields of musical knowledge representation, machine learning of music and software sound synthesis.

Karin Sondergaard
Karin Sondergaard has been working as an independent artist producing most of her works within the institution Boxiganga. The works include performance theatre, video and installation. She is a PhD candidate at CAiiA, in the Planetary Collegium University of Plymouth

Jim Gimzewski and Victoria Vesna
Dr. James K. Gimzewski is Professor of Chemistry in the Department of Chemistry and Biochemistry at the University of California, Los Angeles (UCLA), Member of the Executive Board of the California Nanosystems Institute (CNSI–UCLA) and Co–Director of the Center for Social Interfaces & Networks Advanced Programming Simulations & Environments (SINAPSE), UCLA.

Dr. Victoria Vesna is an artist, professor and chair of the department of Design/ Media Arts at the

UCLA School of the Arts. Her work can be defined as experimental research that resides in between disciplines and technologies. Victoria has exhibited her work in 16 solo exhibitions, over 70 group shows, published 20 papers and gave over 100 invited talks in the last ten years.